BUILDING A SMARTER COMMUNITY

APPLYING GIS

BUILDING A SMARTER COMMUNITY

GIS FOR STATE & LOCAL GOVERNMENT

Edited by
Christopher Thomas
Keith Mann

Esri Press
REDLANDS | CALIFORNIA

Esri Press, 380 New York Street, Redlands, California 92373-8100

Printed in the United States of America
25 24 23   2 3 4 5 6 7 8 9 10

ISBN: 9781589486843
Library of Congress Control Number: 2021936215

For purchasing and distribution options (both domestic and international), please visit esripress.esri.com.

On the cover: Photograph by Bianca Ackermann.

CONTENTS

INTRODUCTION

OBEYING THE LAW, SERVING ON A JURY, AND PAYING TAXES are fundamental aspects of civic responsibility and community stability. However, the greater goal of making a community a better place to live requires citizens and governments to work together to affect change and address larger economic, environmental, and social issues. State and local governments form the front lines of improving the lives of residents by responding to community needs and requests, increasing access to social services, and introducing initiatives and programs that improve the infrastructure, safety, and health of communities within their jurisdictions.

In this book, the word *community* includes all the ways that people are combined within governmental jurisdictions, from small neighborhoods to regional areas (such as counties), to large jurisdictions (such as states and provinces). These communities are inherently entangled with each other through location. For example, a neighborhood exists within a city, which in turn is part of a county that is contained within a state. Governance—the processes that ensure accountability, openness, equity and inclusiveness, and participation—flows between communities.

With smart communities, the word *smart* refers to the use of technology, accurate and timely data, and scientific methods to solve problems. A smart community uses innovative approaches, including location intelligence, to collect and share information and drive good decision-making.

Smart communities don't wait for someone to deem them smart; they take action. One of the most effective actions is to apply geographic information system (GIS) technology to planning and urban design, operations (such as public works), and social issues (such as racial equity and social justice). With GIS, state and local governments make more informed, data-driven decisions that lead to improved outcomes. These outcomes can be shared and replicated in other communities, helping citizens and governments work better together for the common good.

This book is organized in four parts.

Part 1: Planning and urban design

With GIS, city and regional planners and urban designers can create realistic alternatives for comparing different planning and design scenarios, including 2D and 3D visualizations that are more relatable to government officials, businesses, and residents. Planners can combine data about diversity, high density and mixed land uses, open space preservation, housing affordability, and economic viability into their analysis, maps, and presentations to the public.

Part 2: Operational efficiency

GIS makes daily operations—such as responding to citizen requests and tracking the progress of projects— more efficient by improving internal coordination between staff and managers. With GIS visualization and analysis tools, government agencies and departments can combine a wide range of technologies, such as connected sensors, drones, performance dashboards, satellite imagery, mobile devices, and web maps and apps.

Part 3: Data-driven performance

State and local governments use GIS to perform spatial analysis, track resources, and determine whether goals and objectives are being met on time, within budget, and across the communities they serve. Government leaders and managers use spatial data to track hot spots of disease outbreaks, monitor operations outside of the office, manage budgets and capital improvement projects, understand social issues and equity across neighborhoods, and identify areas for environmental and economic restoration.

Part 4: Civic inclusion

Maps and spatial analysis promote a location-centric point of view that is critical to meeting the needs of residents. State and local governments use GIS to engage and collaborate with residents in creative and informative ways, but they also express the value of what's being done—through data—to support all residents. GIS maps and analysis provide individuals and businesses with open data and tools that enhance government transparency, keep the public informed, encourage public participation, and shine a light on racial equity and social justice.

Stories and strategies

Each part includes real-life stories that illustrate how state and local governments use GIS to solve specific problems. Each part concludes with a section about getting started with GIS, which provides ideas, strategies, tools, and suggested actions that government professionals can take to build location intelligence into decision-making and operational workflows.

The stories and strategies encourage the use of GIS to gain a geographic perspective and integrate spatial reasoning into the physical and operational characteristics of government departments and initiatives. The book presents location intelligence as another crucial layer of knowledge that government managers and workers can add to their existing experience and expertise, offering a unique geographic perspective that can be incorporated into daily operations and long-term projects and initiatives.

If location intelligence isn't currently part of a government agency's decision-making processes or considered in daily operational activities, or if it isn't used to improve the lives of its residents, managers can use this book to start developing spatial reasoning skills in those areas. Developing these skills does not require GIS expertise nor does it require government managers and workers to disregard all their experience and knowledge. Spatial reasoning adds another way to think about problem-solving in a real-world context.

HOW TO USE THIS BOOK

THIS BOOK IS DESIGNED TO HELP YOU ADD SPATIAL reasoning to decision processes and operational workflows. It is a guide for taking first steps with GIS and applying locational intelligence to common problems. Using the information from this book can help you create a more collaborative environment within your department and throughout your organization. You can use this book to identify where maps, spatial analysis, and GIS apps can be helpful in your work and then, as a next step, learn more about those resources.

Learn about additional GIS resources for state and local governments by visiting the web page for this book:

go.esri.com/bsc-resources

PART 1

PLANNING AND URBAN DESIGN

WITH GIS, STATE AND LOCAL GOVERNMENTS USE location data to plan for and minimize the impacts and stresses of change. GIS helps planners and urban designers create realistic alternatives that include statistical and spatial analyses for comparing the pros and cons of different scenarios. Interactive mapping and 3D visualizations representing accurate urban and rural landscapes are more relatable to government officials, businesses, and residents than static drawings and other highly stylized renderings. By using GIS to visualize and study proposed transit services, compare capital investment options for a downtown area, find the most effective locations for economic development, or simply reevaluate zoning changes, planners take a forward-looking approach to problem-solving, which leads to more confident actions and better outcomes.

Having a location-centered understanding of a community or place allows planners and urban designers to elevate the value of data through mapping and spatial analysis. Issues such as diversity, high residential density, mixed land uses, open space preservation, housing affordability and scarcity, and economic viability can be measured and investigated together against accurate, lifelike landscapes and neighborhoods. GIS also facilitates open communications with the public, encouraging participation in the planning process and ensuring that transparency and accountability are driving those interactions.

Planners and urban designers use GIS to bridge the divide between what is currently taking place and how people work together to build for the future.

Case studies

Today's state and local government planners are using a mix of digital technologies and strategies to make their communities more sustainable and better places to live. GIS analysis and visualizations help planners and urban designers dig deeper into valuable data and create different scenarios, which can be explored and compared in order to find the best solutions for every project. Using GIS, planners gain a more informed understanding of where negative impacts to the community are reduced and where positive outcomes are most likely.

In the following case studies, city and county planning efforts are highlighted. Planners use GIS to sift through complex data to find the right solutions and to communicate with local officials and the public about what the future might look like.

SHOWING THE COMMUNITY WHAT IS POSSIBLE

Oshkosh, Wisconsin

THE CITY OF OSHKOSH REGULARLY FACES DECISIONS THAT require the community to balance residential and business interests, as well as infrastructure and environmental needs. Elected officials must understand the potential impacts of planning and engineering projects because they're making decisions that affect generations to come, said Kelly Nieforth, the city's economic development services manager. This was especially true when Oshkosh, one of the smallest cities in the nation to boast a Fortune 500 company, overcame initial public disapproval, time constraints, major infrastructure challenges, and more in a quest to retain its largest employer, Oshkosh Corporation, and save hundreds of local jobs.

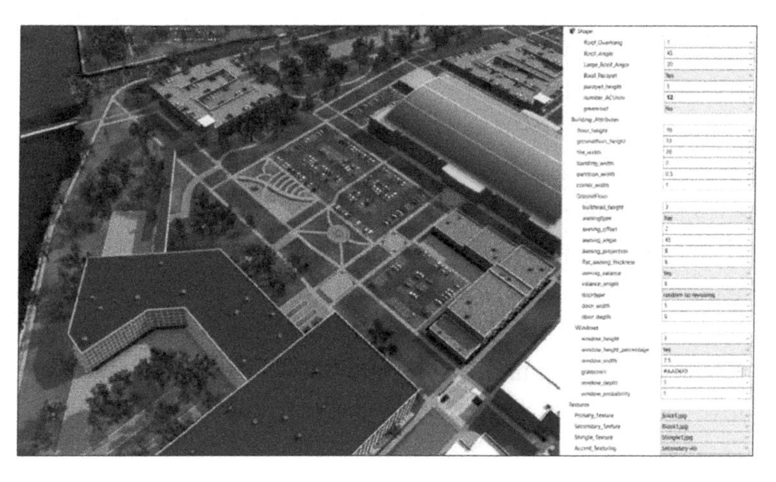

The City of Oshkosh used 3D GIS for planning purposes and for showing the community what's possible. This map is an example of how the city used Esri's CityEngine software to create and share their 3D visualizations.

The city is a longtime user of ArcGIS and Esri tools, especially for planning purposes and for showing the community what's possible, Nieforth said. It's hard for people who aren't in planning or working on projects to visualize that potential. Some people may see only rundown buildings, but others can see a bustling, revitalized area. As part of its proposal to retain Oshkosh Corporation and become the location for the company's new global headquarters, the city used ArcGIS to look at potential sites and settled on offering a portion of the city's municipally owned golf course. Oshkosh also teamed with Esri partner Houseal Lavigne Associates and used CityEngine, an advanced 3D modeling software, to quickly create an interactive and immersive environment for the community and elected officials to weigh in on before proposing that golf course site to Oshkosh Corporation board members. "It helped us put our thoughts in a visual representation" and to have a conversation, Nieforth said.

Not only did the company accept the proposal, but city residents who were initially concerned that a massive office complex would hinder their lakefront views became proponents of the project. In 2018, the city's parks department began work on a plan to turn the portions of the golf course not sold to Oshkosh Corporation into a park. Previously, if people wanted to enjoy the lakefront, they had to pay, said Nieforth, an Oshkosh native. GIS played a vital role in fostering civic engagement and providing an inclusive environment for residents to voice their concerns and to be heard. Nieforth expects that dialogue will continue to flourish as the city looks to tools such as ArcGIS Urban to understand the impacts of zoning codes and to drive conversations with an increasingly informed community.

This story originally appeared as "A 3D Model Provides the Vision to Combat Blight" by Matt Ball, August 30, 2017, on the *Esri Blog*. All images courtesy of the City of Oshkosh unless otherwise noted.

UNDERSTANDING THE IMPACTS OF NEW DEVELOPMENT

Seattle, Washington

WITH THE BOOM OF AMAZON, MICROSOFT, AND THE MANY internet and health care companies that call Seattle home, the city has added 105,000 new residents since 2010 and was the decade's fastest growing US city. Seattle is in the process of updating its 20-year comprehensive plan, which will assess the city's capacity to accommodate that growth. By law, King County is one of several in the state that must determine if they have adequate amounts of residential, commercial, and industrial lands to meet the growth needs. In previous years, planners would use an Access database before plotting that data on a map.

Seattle recently implemented a new building permit system and invested in 3D capabilities to visualize the whole city alongside zoning requirements using ArcGIS Urban software. "For the first time, we are going to use GIS and 3D capabilities to refine our analysis but [also] share it out to the public and to our decision makers," said Jennifer Pettyjohn, a senior planner for the city. "We have quite a large land use code that is a lot of legalese," Pettyjohn said. The city's zoning rules and land-use code are complicated. If printed out, they would create a stack of paper at least 10 inches high. The classifications of zoning codes have all been coded into ArcGIS Urban to improve understanding. The goal is to use the technology to understand where Seattle is now in terms of land use and accommodations for projected growth and to create different scenarios to consider how best to support future residents.

We have had a lot of growth, [and] we expect a lot of growth," Pettyjohn said. "What that means is our planning decisions have

Seattle's zoning rules and land-use code are complicated. If printed out, they would create a stack of paper at least 10 inches high. In this map, the classifications of zoning codes have all been coded into ArcGIS Urban to improve understanding.

to be more transparent. We have to demonstrate that we can look at many different scenarios of how we plan to accommodate the growth." For example, there's a need to understand the impact of new development and what happens to people and jobs when buildings and neighborhoods transform. The impact of gentrification and whether it factors into the growing homelessness problem are of particular concern. Seattle plans to share its technical and methodological advancements with its regional peers to help them compose their buildable lands report. Internally, GIS enables the city to better store its data and provides a single system of record. Not only that, but now planners and developers can turn that data into visualizations. "I've been doing this for 30 years," Pettyjohn said. "I can't believe that I can put out a 3D model on the internet. It's pretty exciting."

This story originally appeared as "Seattle: New Residents, Tech Boom Demand Agile Growth" by Brooks Patrick, February 4, 2020, on the *Esri Blog*. All images courtesy of the City of Seattle unless otherwise noted.

USING GREEN INFRASTRUCTURE TO FUEL SMARTER GROWTH

Richland County, South Carolina

IN 2015, A TWO-DAY STORM CAUSED BY HURRICANE Joaquin swept across South Carolina, leading to devastating flooding, more than a dozen deaths, and the destruction of critical infrastructure. The state recorded $12 billion in losses with up to 160,000 homes damaged.

Richland County was hit worst. This severe weather event dumped two feet of water in some areas of the county, killing nine people and threatening the drinking water. Schools, businesses, and roads remained closed weeks after record rainfall and dam failures occurred.

The historic flood urged decision makers to ask serious questions. Why did flooding occur in this area? What caused the dams to fail? What can we do to prevent this from happening again?

"Flooding is a natural event, but the impacts we experience are caused by humans," said Quinton Epps, director of the Richland County Conservation Department. "We build in the areas that are flood prone. Very often, we change the areas in dramatic ways...that increase flooding impacts. These are things people have known for hundreds of years."

Homes built near bodies of water, for example, are appealing and tend to have higher property values, but they are predisposed to flooding. In Richland County, regulations require developers to build houses that would accommodate the type of flooding that occurs roughly every 100 years, meaning the structures must be built two feet above the base of a potential flood. Although subdivisions

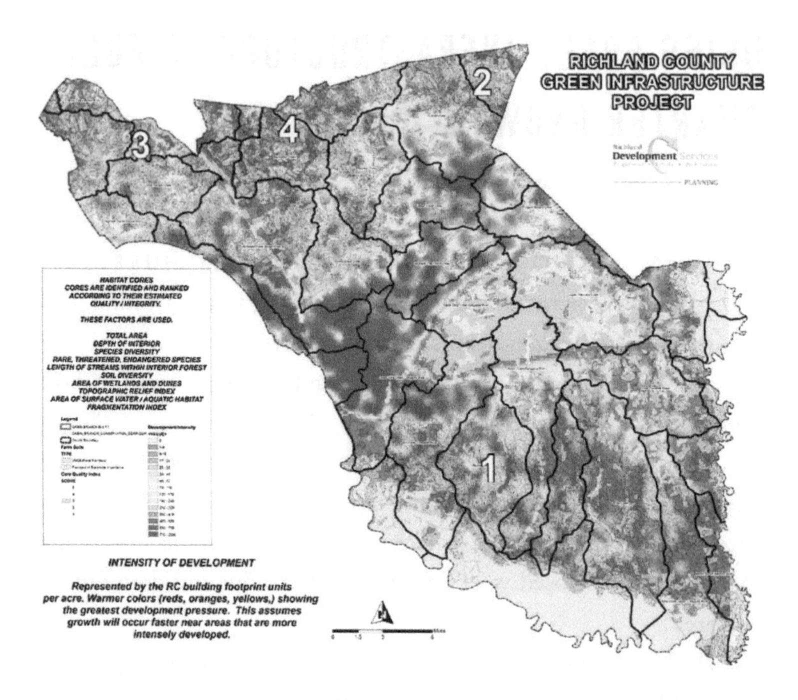

Richland County's planning and development services department used the GIS green infrastructure tool to identify four priority areas in the county that met the criteria of being an intact core, or habitat, at risk of losing its natural assets.

meet current regulations, they are still in danger of being damaged by more severe floods like the county saw in 2015.

As the person in charge of securing properties and funding for conservation efforts, Epps also acquires easements to help avoid disasters like the 2015 flood. He leads projects to connect people with nature as well.

The job is no easy task. Success relies heavily on factors that are out of his control, such as the availability of land, willingness of property owners to sell or donate easements, and political interests that drive funding and development decisions.

After Hurricane Joaquin, Epps advocated to fund easements in flood zones to ensure that the land is never developed. But Richland County allocated its limited grant dollars from the Federal Emergency Management Agency (FEMA) and the US Department of Housing and Urban Development elsewhere—namely, to rebuilding homes and roads.

"The way people look at disaster recovery tells the story of how we don't look at long-term consequences," Epps said. "Do we have money to fix the problems of the last 50 to 100 years of development? No. But we can try to prevent history from repeating itself."

To Epps, most disaster recovery and development strategies focus on rebuilding and extending existing manufactured infrastructure. However, these strategies dismiss a critical component of creating a safe and livable community: green infrastructure.

With Richland County still reeling from the flood, a green infrastructure model could do more than connect people with nature; it could also help win executive buy-in for conservation projects that safeguard wildlife habitats, protect people and property from future harm, and boost the economy. By using GIS to assess all the county's features that the community deems valuable, Richland could explain the importance of green infrastructure in a more scientific way.

"We could provide a rational basis for conservation efforts not tied to politics or opinions," Epps said.

Richland County's planning and development services department used the GIS green infrastructure tool to identify four priority areas in the county that met the criteria of being an intact core, or habitat, at risk of losing its natural assets.

He was inspired to formulate a green infrastructure plan that would enable the county to plan smarter, and he knew one person would be the key to developing the plan—Brenda Carter, GIS manager for the planning and development services department at Richland County.

"Reading through everything made me realize that this was something really, really important for all counties and for all people—especially after our county had just suffered a great flood," Carter said. "I started seeing the connections as to why the flood could have happened and why we needed to do something about it."

Thus, the planning and conservation departments formed a partnership to bring green infrastructure planning to Richland County. Carter outlined goals and developed the entire green infrastructure blueprint. She began by identifying four goals:

- Improve water quality by providing a buffer to help prevent runoff and erosion and reduce pollutants.
- Maintain forested land cover to facilitate recharging groundwater aquifers for drinking water.
- Preserve and promote natural resource-based recreation, such as hiking, bird watching, hunting, and fishing.
- Conserve community character and heritage by protecting a historic landscape.

Once Carter had the essential data layers in place—including existing county GIS layers such as conservation easements, address points, and zoning—she used tools from Esri to create asset maps and maps of intact habitats, also referred to as cores. Carter employed spatial analysis using ArcGIS to conduct landscape analyses, assess fragmentation and risks, develop a core quality index, and prioritize opportunities.

"The first step to mapping the intact habitat cores is identifying the location and shape of habitat cores," Carter said. "The second

In this map, green infrastructure analysis shows areas identified as targets for land conservation. Yellow lines show the boundary of the two conservation easement properties, and red lines enclose a riparian area that cuts both easements.

step is ranking the cores based on their ecological integrity using the best available data and science."

With the green infrastructure tool, Carter identified four priority areas in the county that comprise unique rural lands and waters. Each of these locations met the criteria of an intact core at risk of losing its natural assets. As part of the green infrastructure plan, Carter and Epps also identified potential projects to protect the cores—such as purchasing easements to preserve certain streams—and made clear the benefits of implementing them.

"I've been doing environmental work for 30 years," Epps said. "I had a picture of what the green infrastructure model would look like and, in the end, it didn't resemble what I had in mind. But that's what [is] cool—Brenda used scientific data and tools to set the strategy. It's not just about picking out what we think we should protect; it's scientifically supported."

The task force presented the green infrastructure plan and maps to the Richland County Conservation Commission, a group of 11 members appointed by the county council to implement conservation goals. The commission was impressed.

"Science proved what they were thinking all along," Carter said. "Now they have scientific evidence to prove which areas need protection and restoration."

The team presented the information to the county council as well, to resolve issues with zoning and the county's comprehensive plan. And the county's land administration department rewrote the land development code, keeping the priority cores top of mind.

"We're not going to write code to keep developers from [building new homes]," said Carroll Williamson, land development administrator with Richland County. But the "new codes will guide smarter development that will be better for the land, people's investments, and our county in the future. We used to tell developers that green infrastructure was a 'nice to have' feature. But if we can say scientifically that green infrastructure is critical to our well-being, then it takes on much greater significance."

Moving forward, the team will continue the project and identify additional priority cores throughout the county's council areas. They're excited about the possibilities that the green infrastructure plan can bring to the county, including for the local economy. Helping people—especially county executives and developers—understand the benefits of green infrastructure will be critical.

Epps explains the green infrastructure planning is not simply about conservation but a way to provide a more sustainable and resilient community and to be better prepared for the next extreme flood. For Carter, green infrastructure planning means using her craft to enhance quality of life for all of Richland County for years to come.

We want to preserve the natural resources that we have in the county so we can protect our quality of life, says Carter. "We want to grow, but we want to have smart growth."

This story originally appeared as "Green Infrastructure Plan Fuels Smarter Growth in Richland County" in the Winter 2017 issue of *ArcNews*. All images courtesy of Richland County unless otherwise noted.

UNCOVERING THE VISUAL PATTERNS OF AFFORDABLE HOUSING

Honolulu, Hawaii

ONLINE HOSTING SITES, SUCH AS AIRBNB AND VRBO, DIDN'T set out to disrupt local real estate markets. However, research shows short-term vacation rentals are becoming a major driver of rising rents and home prices in cities around the world.

In Hawaii, and Honolulu in particular, affordable housing has been hard to find for a long time. To combat this scarcity, and to balance the demands of residents versus tourists, the Hawaii legislature recently passed a bill, awaiting the governor's signature, that requires hosts of short-term rentals to collect and pay lodging taxes. The Honolulu City Council just passed a bill that takes the more severe step of restricting the number of short-term rental units and fining violators up to $10,000 per day.

The state tax could raise considerable revenue as Honolulu is home to 800 legal vacation rentals and an estimated 8,000 illegal units. Restricting the number of short-term rentals should return some of these units to use by locals. To gain an understanding on these changes and how they impact residents, city analysts and planners use GIS technology. In Honolulu, GIS mapping and databases are available through the City and County of Honolulu's Land Information System (HoLIS).

"Honolulu (the City and County of Honolulu) has begun to look at changing the Land Use Ordinance (zoning codes) to address affordability," said Ken Schmidt, GIS administrator for the City and County of Honolulu. "GIS helps quantify how this problem came to be and allows us to visualize ways we might address it."

In this map, Honolulu planners use GIS to visualize a zoning change and the potential impact to the community.

According to a 2015 report, 66,000 housing units will be needed in Hawaii by 2025 to meet demand, with nearly 26,000 of those dwellings required in Honolulu. A number of proposed zoning changes aim to address this looming housing deficit.

Proposals include restrictions on the square footage of residential units to combat monster homes; an easing of height restrictions on low-rise apartments to allow five-story walkups rather than the existing three-story limits; and a proposal allowing homeowners to build and rent accessory dwelling units.

The Honolulu Department of Planning and Permitting used 3D visualization tools to examine and visualize the proposed low-rise apartment zoning change in the neighborhood of Mo'ili'ili. Honolulu worked with Esri, piloting the capabilities of ArcGIS Urban and CityEngine. "We've been working to demonstrate to planners what the changes to our zoning code would look like in 3D," Schmidt said. "Our planners' eyes got really wide when they realized that they could quickly and easily change values in the zoning code and see what the difference in development would be."

Metropolitan Honolulu is the fourth most densely populated area in the US, with traffic that often ranks among the worst in the country. The constraints of the island, with its narrow coastal plains, steep slopes, and cross-island canyon connections cause much of this congestion.

The 20-mile, 21-station Honolulu High-Capacity Transit Corridor Project aims to help alleviate the problem by introducing elevated rail to the region. The rail corridor, under construction now, would move 8,000 people per hour and give Honolulu residents an opportunity to rethink development patterns. The city has already embraced Transit-Oriented Development (TOD) as a strategy to increase the density of housing, jobs, and services around rail stations.

"We've been working with the TOD division to look at changes in zoning laws to provide incentives for development within a half mile of station locations," Schmidt said. Schmidt and his team created a participatory environment that allows government, business, and community stakeholders to visualize TOD redevelopment scenarios and rezoning proposals.

As Honolulu begins the build-out of the new rail line and associated development, officials are looking to improve the city's permitting process. "Anybody who tries to get a building permit usually has some challenges," Schmidt said. "That's primarily due to the amount of volume as we have from 15,000 to 20,000 permit applications per year. Anything we can do to help streamline that process or make it easier is one of our major objectives."

The city was an early permitting innovator with a system called HONLine that allows homeowners to quickly obtain permits for such things as fencing, solar panels, water heater replacements, or other minor upgrades to residential property. "GIS plays an important role because we have to check to see whether that property might be in a flood zone, along a coastline, or how it's zoned," Schmidt said. "The integration of the permitting system with our GIS was critical in our ability to be able to put those permits online."

Next, Schmidt and his team hope to tackle improved business workflows for new development by incorporating digital building information models (BIM) within GIS as part of the permit review process. Digital BIM files would replace the paper plans in use today. A new all-digital workflow would provide a means to streamline building code assessments and automate the detection and notification of any conflicts.

In this map, planners can model and analyze different scenarios based on building heights and locations.

While ongoing development and redevelopment in Honolulu have considerable impact on the local economy, GIS provides the means to quantify it. "Looking through our data, we can see 0.2 percent of the total number of building permits accounted for more than 10 percent of the total value of all building permits," Schmidt said. "That small number of permits reflects 33 new apartment complexes built in the last year."

The impact of new apartments, on both the affordable housing issue and the overall city economy, is one of the factors driving the city to recently adopt the proposed walk-up apartment zoning change. Many more ideas are coming forth to deal with the state's housing crisis. "A variety of different creative and innovative—and somewhat haphazard—proposals are coming forth," Schmidt said. "Each proposal needs to be analyzed using GIS to see potential impacts and to communicate the results to the public."

This story originally appeared as "Honolulu Planners Visualize Housing Patterns with an Eye on Affordability" by Brooks Patrick, June 25, 2019, on the *Esri Blog*. All images courtesy of the City of Honolulu unless otherwise noted.

GETTING STARTED WITH GIS

ISSUES SUCH AS HOUSING AVAILABILITY, SUSTAINABILITY goals, and economic changes are compelling state, regional, and city governments to better plan for the future. GIS allows planners and design professionals to collaborate across teams using maps, spatial analysis, and 2D and 3D applications. GIS also supports scenario planning and impact assessment.

GIS offers a way to view, analyze, and share planning projects with internal stakeholders, private partners, and the public. With GIS, planners and urban designers can create a detailed, illustrated project, from beginning to end, including the evaluation of different scenarios and spatial measurements of the impacts to the community.

Planners and urban designers can use GIS to see data in new ways, bringing together data from many sources and visualizing all the relevant information in one place. When planners are working with spatial data, they can start asking spatial questions.

How do I use GIS to visualize planning projects?

With GIS, planners and urban designers can create, track, and review development plans with a digital twin (a virtual and editable image) of project areas. The digital twin provides a direct connection between the physical area of the project and all the information, assets, guidelines, and policies within one unified, digital model. For example, building information modeling (BIM), areal and satellite imagery, and 3D information can be incorporated and investigated throughout the project review process, giving planners and developers a common view of development guidelines.

How do I make 3D scenarios a part of the planning process?

GIS enables planners to create, edit, and manage interactive 3D environments. GIS 3D scenarios simulate what proposed planning areas might look like, given different criteria, within the context of the larger picture. With GIS 3D, plans include realistic buildings and structures alongside zoning and land use designations, rights-of-way and setbacks, and details of integrated infrastructure for an all-inclusive view of current or future development efforts. A 3D scenario-driven process makes land-use changes, zoning updates, and evaluation of development proposals faster, which in turn, improves communication and speeds up policy-making decisions.

How does GIS help me guide change and measure impacts?

GIS allows planners and urban designers to directly join essential data from many different sources within planning maps and analysis, including databases and repositories. Factors such as population change, economic growth, and housing availability provide planners with an all-inclusive view of project areas, which leads to a more comprehensive understanding of all the factors affecting a comprehensive planning process. GIS enables spatial measurements and statistics, such as hot-spot analysis and sightline analysis, that help planners guide people through the impacts of proposed changes within and around a project.

How can I use GIS to increase stakeholder participation?

GIS applications increase the reach of planning professionals through sharing interactive, digital maps. Online maps open access to new audiences, often boosting access to key demographic groups and stakeholders. GIS-based applications allow planners to collect online comments and survey affected populations, as well as clearly communicate development proposals in public and internal meetings and on the web. 3D GIS provides public reviewers and commenters with an interactive view of alternatives that include simulations and analysis of building heights and shadows, distances to stores, and proximity to bus stops, subway or light rail stops, and parking. With location-based visualizations and metrics, residents and business leaders get an inside track on new information affecting their interests, leading to more participation in the planning process.

Using GIS

There are two ways to get started with GIS: hands-on learning and using ArcGIS Solutions.

Hands-on learning

Hands-on learning will strengthen your understanding of GIS and how it can be used to improve planning and urban design. A good place to start is with Learn ArcGIS, an online resource that introduces GIS using real problems and scenarios. Learn ArcGIS lessons will help you understand how planning and 3D maps are represented and learn more about the following:

- ArcGIS Urban and other GIS tools for creating 3D planning maps and analysis, such as creating alternative scenarios by changing zoning and parcel information
- Using GIS to encourage citizen participation in the planning process and improve public service using data from GIS survey forms and public-facing portals such as ArcGIS Hub

- Creating a greenfield plan that helps planners work with land developers, neighboring jurisdictions, and staff to ensure environmental sustainability
- Studying the per capita environmental impacts of industrial facilities and meeting standards for monitoring and regulating the discharge of toxic chemicals into the environment
- Communicating planning initiatives and engaging the public with maps, images, and text that break down the big ideas shaping their communities

Learn about additional GIS resources related to state and local government planning and urban design by visiting the web page for this book at **go.esri.com/bsc-resources.**

ArcGIS Solutions

ArcGIS Solutions is a collection of focused maps and apps that help address challenges in your organization. As part of the Esri Geospatial Cloud, solutions work with your data and are designed to improve operations and gain new insights. For example, you can use one of the following methods:

Create and publish 3D basemaps

Create a collection of high-quality 3D basemaps using existing data such as lidar, building footprints, and underground utilities to use in desktop, mobile, and web mapping applications.

Integrate GIS with capital improvement planning

Make mapping and spatial analysis part of capital improvement planning, including defining projects plans, coordinating project schedules with internal and external agencies, and organizing a project portfolio that can be shared with the public and other key stakeholders.

Share incentive zones with businesses

Use GIS maps and apps to promote community growth opportunities in targeted areas by sharing incentive zones with business owners and corporations to encourage new businesses or the expansion of existing businesses in a community.

Conduct homeless point-in-time counts

Conduct location-based annual counts of homeless persons who are sheltered in transitional housing, emergency shelters, and safe havens and quickly summarize the extent of homelessness and streamline requests for federal funding.

Takeaways

State and local government planners and urban designers use location data to express the value of what's being planned and drive decision-making through mapping, spatial analysis, and evaluating the impacts on people, businesses, and the environment. By integrating GIS and spatial thinking into comprehensive planning efforts, planners and urban designers take a forward-looking approach to problem-solving, which leads to more confident actions and better outcomes.

In the Oshkosh story, city planners and officials used 3D GIS to overcome initial public disapproval and major infrastructure challenges in order to retain its largest employer and save hundreds of local jobs. In the Seattle story, the city used different GIS-based scenarios to plan for the impact of new development and accommodate the community growth. In the Richland County story, planners and conservationists applied a green infrastructure approach that helped win executive buy-in for conservation projects that safeguard wildlife habitats, protect people and property from future harm, and boost the economy. Finally, in the Honolulu story, city analysts and planners used GIS to address the inequities of housing scarcity and the city's housing crisis and formulate location-specific zoning changes that streamlined building code assessments by automating the building permit conflict detection and notifications.

This section also provided recommendations for learning and quickly applying GIS tools in planning projects.

Next, you will learn about using GIS to improve the efficiency of daily work and reduce the costs and time needed to perform continuous operations and maintain critical services.

PART 2

OPERATIONAL EFFICIENCY

PRODUCTIVITY AND EFFICIENCY ARE ESSENTIAL QUALITIES in any successful organization. While state and local governments generally operate differently than private companies, they still adhere to similar principles, such as financial and fiscal accountability, achieving stated goals and objectives, and meeting the needs of their customers. The key difference in how a government operates, as opposed to a corporation, involves making its work more open and reviewable by the general public.

Attaining operational efficiency can take many forms, but it usually includes finding new ways of working that enable departmental staff to optimize the management and performance of internal operations and constituent services. New workflows can reduce the cost and time it takes to deliver existing services and make new services possible. Government agencies gain efficiency when they improve how they collect, share, and analyze location-based data and integrate GIS with operational decision-making.

With GIS visualization and analysis tools, government agencies and departments can combine a wide range of technologies—such as connected sensors, drones, performance dashboards, satellite imagery, mobile devices, and apps—into operational decisions and work. GIS helps state and local governments become more efficient and effective in the following ways:

- Transition away from cumbersome legacy processes and inefficient paper-based workflows by using GIS maps and mobile apps to create asset inventories and perform inspections.

- Improve data collection abilities and workforce mobility using GIS apps on smart devices, drones, and sensors to increase data timeliness and accuracy while doing maintenance projects and responding to citizen complaints.

- Accomplish faster data analysis and mapping that supports decision-making and mobile workers in daily operations.

- Increase response effectiveness for citizen requests and emergencies using real-time GIS dashboards to monitor and share performance metrics and visualize trends.

- Create better opportunities for productive participation within the community by sharing GIS maps, dashboards, and reports with civic leaders, business owners, and the public.

Case studies

Operational efficiency is the cornerstone for measuring how well state and local governments are performing. As a result, governments are exploring new technologies, including GIS, to create cost-effective programs and initiatives that reimagine the way they provide efficient services to residents. Data is at the center of these innovative approaches. In the following case studies, you will see how governments and agencies reinvent the flow of data using GIS maps and apps, machine learning, augmented reality, and powerful open-source libraries to perform more efficiently in day-to-day operations and large projects.

PRIORITIZING REPAIRS ON ROADWAYS WITH GIS

District of Columbia, United States

IN 1791, PRESIDENT GEORGE WASHINGTON COMMISSIONED French engineer Pierre Charles L'Enfant to plan and design the District of Columbia, the permanent capital of the recently founded United States of America. L'Enfant envisioned important buildings and monuments occupying strategic locations based on changes in elevation and the contours of local waterways. He took inspiration from European cities, with the National Mall patterned in part after the Champs-Élysées in Paris, France, and streets radiating out from the city center to resemble Karlsruhe, Germany. Today, the capital's roadway network is maintained primarily by the District Department of Transportation (DDOT) and includes 1,100 miles of public roads, both local and federal; 1,495 miles of sidewalks; and 358 miles of alleyways.

Because of heavy traffic throughout the year and adverse weather conditions during winter months, the window to repair and maintain the capital's roadway network is small. It starts each spring and continues into the fall, requiring a concerted effort to accomplish all the necessary work.

In 2018, Mayor Muriel Bowser announced the launch of the PaveDC Paving Plan, a comprehensive initiative to fix all roads rated in poor condition in Washington, DC, by 2024. PaveDC has four priorities: road rehabilitation, road maintenance, alley repair and reconstruction, and sidewalk reconstruction.

"For the PaveDC initiative, we first created the PaveDC Asset Repair Prioritization Model," said Dr. Ting Ma, performance manager at DDOT. "We began by assembling three different datasets.

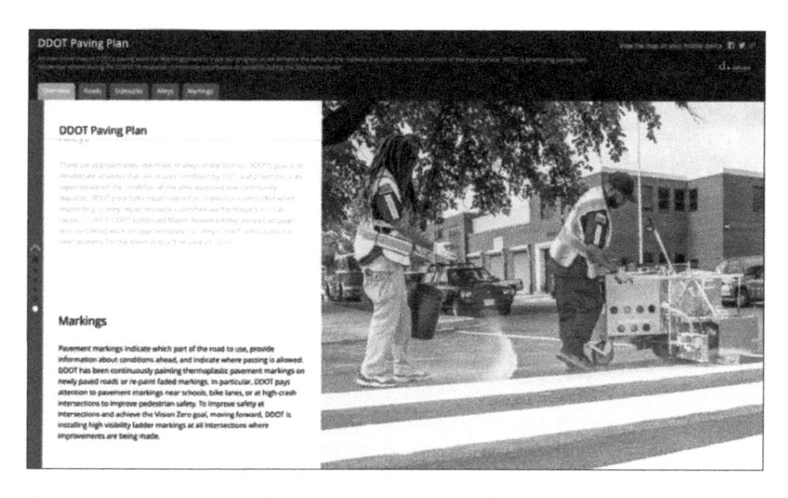

The District Department of Transportation (DDOT) Paving Plan website is an interactive and real-time paving plan communication tool that keeps the public informed about road, sidewalk, and alley projects.

The primary dataset comes from our annual pavement condition index. This is created by surveying every road in the district. Then, the roads are assigned a value from 0 to 100 based on their condition—how many potholes, how many cracks, and so on. Those roads with the lowest scores need the greatest attention."

DDOT derives another dataset from its 311 service call request line, which residents use to report nonemergency roadway problems. The calls are tabulated according to location and the type of repair requested and then compiled into a dataset. In addition, DDOT has an Americans with Disabilities Act (ADA) compliance dataset to ensure that people of all abilities can travel safely around the city.

"These three datasets are compiled and then an algorithm is applied to the data that creates a new index, called the Roadway Repair Index," said Ma. "It is used to rank the need for repair of each roadway in the district and prioritize the repair and construction

DDOT built an app that supports collaborative editing and data verification and has street view images from Cyclomedia.

work. Using these three separate datasets provides us with a more comprehensive and holistic evaluation of roadway conditions."

PaveDC enables DDOT to pave more roads more quickly. In 2018, for example, the department paved 55 miles of roadway, whereas in 2019, it paved more than 100 miles of roadway—the most DDOT has ever done in a single year.

DDOT is using the ArcGIS Roads and Highways extension to build its digital roadway network and maintain the data. For the PaveDC analyses, Ma uses a segmentation tool and geoprocessing tools in ArcGIS Pro to aggregate the data from three sources down to the block level. From there, she can visualize the data and apply weights to test different cost estimates for road repairs. This data visualization is used to balance road repairs in all eight wards to ensure that DDOT maintains roads equally throughout the district.

In addition, DDOT used ArcGIS Web AppBuilder to create an app that supports collaborative editing and data verification and has street view images from Esri partner Cyclomedia. "Our engineers and inspectors use this app to determine whether or not there is a conflict on the roads we intend to repair with underground utility work being performed in the same area," said Ma.

To keep the public informed about its work, DDOT created and launched the DDOT Paving Plan website in 2018. It's the agency's first interactive and real-time paving plan communication tool for the public to explore asset repair and work plans and track the status of projects. By presenting an overview of the various services PaveDC provides, the website informs users about what to expect before and during construction. Users can find specific information on roadway paving, sidewalk and alley repair, and pavement-marking improvements on four separate dashboards. The dashboards are interactive, allowing users to click assets to find information about them or review specific wards or the advisory neighborhood commissions (ANCs) they are interested in. The website uses several ArcGIS products, including ArcGIS StoryMaps, ArcGIS Dashboards, and

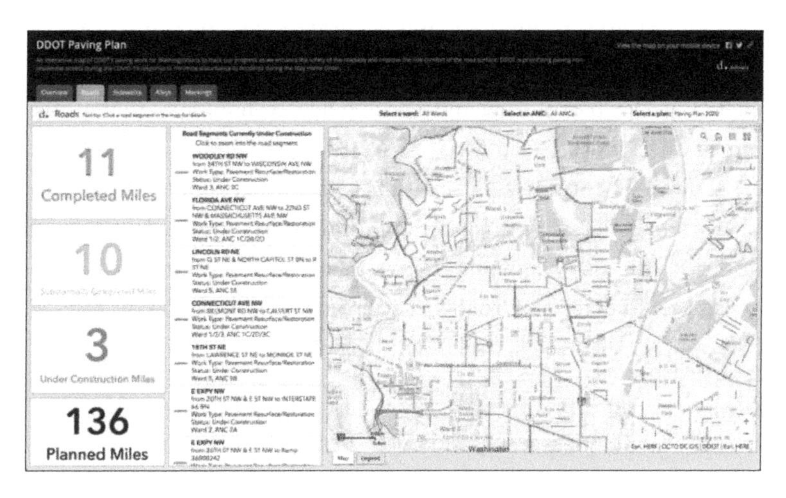

Using four dashboards set up in an ArcGIS StoryMaps app, website visitors can see how many road, sidewalk, and alley repair projects are either in progress or already finished.

ArcGIS Online maps, to convey details about the projects DDOT is working on. It shows the status of the repair work, how many miles of roadway are currently being repaired, and how many miles have been completed. To communicate more effectively and directly with residents, DDOT posts PaveDC updates to its Twitter account as well, using the hashtag #PaveDC to connect the posts.

"The website allows us to provide greater transparency and engagement with the general public," said Ma. "Our residents love our PaveDC website. Since its launch, it has received more than 30,000 visits." The website also keeps various divisions within the transportation department up to date on progress, which provides an incentive to continue doing innovative work, according to Ma.

"We use it to present a status report on the work we have accomplished during the past week," she said. "This takes place at our weekly meetings with the executive team. So internally, PaveDC is also a performance management tool." PaveDC provides a good example of how technology and accurate data can transform the way transportation agencies manage their infrastructure assets. "At DDOT, we keep exploring and learning new technologies to improve our PaveDC work," said Ma.

DDOT is currently automating its geoprocessing and data prioritization process using ArcPy and other open-source data science libraries. It is also beginning to investigate the use of artificial intelligence (AI) to determine roadway repair needs more quickly.

This story originally appeared as "GIS Helps Prioritize Repairs for Washington, DC's Heavily Trafficked Roadways" in the Summer 2020 issue of *ArcNews*. All images courtesy of the District of Columbia unless otherwise noted.

STRENGTHENING DATA COLLECTION WORKFLOWS

City of Salinas, California

WHEN DESCRIBING THE RELATIONSHIP BETWEEN WATER and the Salinas Valley, John Steinbeck said in his book, *East of Eden* (1952), "I have spoken of the rich years when the rainfall was plentiful. But there were dry years too, and they put a terror on the valley."

While California's relationship with rainfall has always been prickly, water management has changed in the valley since Steinbeck's time—arguably becoming even more critical to sustaining life in the arid American West.

Located 60 miles south of Silicon Valley, the City of Salinas, California, is surrounded by agricultural fields yet is one of the state's 50 most densely populated cities. For its water supply, Salinas relies exclusively on local groundwater. But because of decades of over pumping, that water source is now threatened by saltwater intrusion. Additionally, due to the high density of impervious groundcover in Salinas, the volume of stormwater runoff that leaves the city each year is excessive, amounting to a missed opportunity for the city to use this valuable resource to recharge its groundwater and supplement local water supplies. This urban runoff also carries pollutants and litter into local waterways—and every acre of land in Salinas drains into a stream that the Environmental Protection Agency (EPA) has designated as impaired.

To radically change how Salinas manages its stormwater, the city's public works department has teamed up with Esri startup partner 2NDNATURE to better leverage ArcGIS to collect smarter data and streamline the workflows that reinforce its regulatory reporting requirements.

Typically, Salinas's stormwater manager spends five months each year gathering data from 12 different city departments to comply with reporting requirements. For the 2016–2017 fiscal year, the city's report to the Regional Water Quality Control Board exceeded 2,800 pages of narratives, lists, tables, and static maps.

City staff were frustrated by the inefficiency of this process and wanted to create actionable information to guide and justify their stormwater management decisions. In January 2017, the City of Salinas started using 2NFORM, a suite of software from 2NDNATURE that works in step with ArcGIS Online, to help public works departments conduct inspections and process data in compliance with their National Pollutant Discharge Elimination System (NPDES) permits—regulations that stem from the federal Clean Water Act of 1972 and require cities to protect local waterways.

The solution is designed to make data collection and management more efficient and to effectively communicate the benefits of its stormwater compliance program within and across city agencies. The fit-for-purpose software integrates topography with city storm drain networks, linking urban landscapes to the waterways into which they drain. This hydrographic foundation empowers Salinas's stormwater managers to apply a watershed context to all their stormwater decisions and regulatory reporting.

One key feature of 2NFORM is that it enables city staff to monitor and evaluate best management practices (BMPs), which are various activities municipalities undertake to control the effects the urban landscape has on water resources. In Salinas, the city and private landowners implement various BMPs—such as street sweeping, litter reduction programs, pollution prevention initiatives, and green infrastructure development—to either control runoff and pollutants at the source or to intercept, capture, and remove pollutants from stormwater runoff.

As a result, Salinas is building a stormwater asset database. City staff catalog BMPs by location and type so that public works can see exactly where they are. While permit regulations require cities to periodically inspect public and private BMPs to demonstrate that they are working, there is typically no uniformity to these inspections, and they are not grounded in geospatial-based record-keeping systems. So cities don't always know what improvements need to be made or where.

Now, city staff employ custom-built rapid assessment methods (RAMs) to generate objective, standardized, and repeatable evaluations of how BMPs are functioning. The RAMs can be conducted using both ArcGIS Collector and ArcGIS Survey123 to record uniform data about each BMP via standardized inspections.

To evaluate the effectiveness of litter controls—such as street sweeping, education, outreach, and adopt-a-street programs—two employees now drive inspection routes with continuous collection (or the "breadcrumb trail" feature) enabled in ArcGIS Collector. This allows them to trace their path while recording how much litter is on the route, which reduces the time it takes to both do inspections and input data manually. Evaluating a bioretention system—like a rain garden that traps rain runoff in a vegetated area to absorb some of the water and release cleaner runoff into the stormwater system—is now completed by public works employees using inspection forms in ArcGIS Survey123 that make assessments simple, repeatable, and rapid. Additionally, anyone in Salinas can conduct litter surveys using the Survey123 trash inspection form. After downloading the form, they can quickly and easily document how much litter there is, where, and when they see it.

All this inspection information feeds directly into 2NFORM, which enables the public works department to conduct additional geospatial analytics and evaluate where it needs to focus maintenance efforts or make improvements.

Beyond stormwater asset management, staff can model stormwater runoff and pollutant loading. Employing publicly available datasets—such as the Natural Resources Conservation Service's (NRCS) soils survey, the United States Geological Survey's (USGS) impervious land-cover data, and local precipitation data—Salinas can estimate the average amount of stormwater runoff and pollutants that get delivered to every local waterway each year.

Once these baseline estimates are calculated, the inspection results are used to evaluate whether the private and public BMPs implemented throughout the city are reducing runoff and pollutant loading. This gives Salinas the ability to track how well its stormwater initiatives and methods are working. It also helps the city actively manage locations and solutions that are not functioning appropriately so it can better protect and improve local waterways.

To demonstrate the benefits of harvesting stormwater, 2NDNATURE created an Esri Story Maps app called CA Stormwater Opportunity that shows the average annual stormwater runoff for communities all over California.

Now that Salinas's public works department can visualize the spatial patterns of stormwater runoff and pollutants, the city can communicate program priorities and progress among departments and to the public. Mapped results allow anyone to see where the city has achieved its runoff and pollutant control objectives, as well as where more improvements are needed.

All the map-based tools available from 2NDNATURE also give Salinas a standardized way to complete its annual compliance reporting and ensure that it is actually implementing effective actions that

In this story by 2NDNATURE, California stormwater opportunities are depicted and quantified for cities in the central coast region. The map is focused on the City of Salinas.

reduce excess stormwater runoff, litter, and other pollutants from ending up in its waterways. Furthermore, public works staff are saving hundreds of hours collecting and managing the data they need to comply with regulations.

By leveraging the web-based infrastructure of ArcGIS Online, science-based stormwater information is now more accessible to many more people. This empowers cities to make smart decisions about stormwater that can reduce the impact of urban development and restore hydrologic function to urban watersheds.

This story originally appeared as "Startup Takes on Stormwater Management—and Salinas Gains Efficiency" in the Summer 2018 issue of *ArcNews*. All images courtesy of 2NDNATURE and the City of Salinas unless otherwise noted.

MODERNIZING WASTEWATER MANAGEMENT

Ontwa Township, Michigan

ONTWA TOWNSHIP, LOCATED IN THE SOUTHWEST CORNER of Michigan along the Indiana state line, serves 6,000 residents and covers 21 square miles. The township has several large lakes, including the home-lined shores of Eagle Lake, a community centerpiece.

Recently, ruptures in the town's wastewater network embarrassed the community and resulted in costly fines, including $60,000 in penalties both for spillage and failure to follow protocol when sludge gushed into the water and onto the land.

"Things break down because we don't have a good alarm system," explained Beth Westfall, a member of the Ontwa Waste Water Board.

Like many small townships, Ontwa's asset information spread over many disconnected software applications, databases, and systems. Operations staff couldn't find the sewer management data they needed. When they did, it was often wrong and outdated.

Recognizing the growing challenges of Michigan's infrastructure, the Michigan Department of Environmental Quality (MDEQ) allocated $450 million to modernize the state's water networks and treatment plants. Called the Stormwater, Asset Management, and Wastewater (SAW) Grant Program, Michigan metes out funds to help communities like Ontwa struggling to manage assets and respond to crises intelligently. Ontwa applied and received a SAW grant to help solve its problem.

The city's wastewater treatment provider, Elkhart, repeatedly saw high levels of hydrogen sulfide gas in the sewers. Hydrogen sulfide in the pipes converts to sulfuric acid, and over time, the acid compromised the porous concrete vaults containing the apparatus that controlled flow and pressure. Eventually, system failures led to a spillage due to a collapsed force main.

After spending more than $100,000 on new pumps, valves, and chemical additives to address the hydrogen sulfide issue, a larger challenge remained: how to better manage the physical network and prevent incidents from occurring. All agreed that being reactive had proved expensive for the community.

Ontwa worked closely with Wightman, a leading engineering firm, to create a "digital twin" of its wastewater network using a GIS to capture network details. With that foundation, the township could then track sewer maintenance efforts and meet protocols should another rupture occur.

Like all distribution systems, wastewater infrastructure consists of many component assets, not just valves and pipes. To prevent future failures, officials recognized the township needed to account for every part by knowing its exact location and condition.

The new asset management system moved Ontwa from outdated manual processes to modern automated workflows.

"What we are really excited about is the ability to schedule, track, and monitor routine maintenance programs, and to have a complete reporting support system," said Ray Galovich, project manager for Infrastructure Alternatives, the firm that provides operational services to Ontwa Township.

The GIS-powered asset management system helps Ontwa answer fundamental questions to prevent future failures:

- What do we own?
- Where is it located?
- What condition is it in?
- What is the remaining useful life?

- What are the most critical assets?
- Where was maintenance performed and what was done?
- Where are improvements needed?

Maintenance staff now use a mobile app that displays an up-to-date map of the water system. With it, staff can view assigned work orders, provide updates based on tasks completed, and access service order and preventive maintenance plans. The updated system also supports the staff's ability to present information to managers and township officials, strengthening Ontwa's commitment to transparency.

With the ability to closely monitor the condition of all assets, Ontwa has moved from reactionary to preventive sewer maintenance—a primary goal of any large-scale digital transformation. Next, the department hopes to integrate more sensors and real-time data feeds along the network to know remotely how the system is performing.

This story originally appeared as "A Michigan Township Modernizes Sewer Management" by Keith Mann and Matt Ball, February 22, 2018, on the *Esri Blog*. All images courtesy of Ontwa Township unless otherwise noted.

IMPROVING OPERATIONS WITH MOBILE GIS

City of Santa Barbara, California

THE CITY OF SANTA BARBARA, CALIFORNIA, LOCATED roughly 90 miles north of Los Angeles, supplies water to approximately 95,000 residents. Given California's proneness to droughts, state water suppliers face operational challenges surrounding conservation, water-loss prevention, and system maintenance.

In 2014, Santa Barbara's Public Works Department team determined that they needed to modernize and optimize their existing water system. To do so, they set forth an ambitious goal of replacing all 27,500 of the city's water meters within five years.

"Instead of just doing a regular meter replacement based on age, we decided to handle this task comprehensively to better set ourselves up for the future," said Theresa Lancy, water distribution supervisor at the City of Santa Barbara. The city needed to modernize its legacy system and existing workflow for field data collection and analysis of its large network of water meters. The meter replacement project—containing an ambitious goal of replacing all 27,500 of the city's water meters—presented an ideal opportunity for the city to strengthen the accuracy of its location data.

In its legacy system, the real-time location of all meters was not adequately captured using a GPS and did not provide the level of accuracy required to sufficiently identify individual meters. This was due in part to the meters being directly correlated with a parcel. There was uncertainty in the data's accuracy when multiple meters were present in a parcel.

"We were using the parcel number to get a general idea of the location of our meters," said Lancy. "But in circumstances where multiple water meters are located on a single parcel, knowing which meter was associated with a particular location was especially difficult."

Since the city's water meters were often located within one foot of each other, the city's team needed a new solution that would provide a higher level of accuracy. As proof of concept, they borrowed a GPS receiver from a sister department. The receiver used a mobile base station to provide differential corrections, but the station needed to be manually set up and broken down. This took an hour of fieldwork every day, slowing down the meter location collection. The city not only wanted highly accurate location data but also needed a faster data collection workflow.

The city used Esri's ArcGIS suite of mobile applications to help meet its immediate need for an improved mobile data collection workflow. ArcGIS enabled the mobile workforce to remotely edit and utilize updated web maps simultaneously. Both mobile and office staff now have a shared view of the water meter data. Santa Barbara's mobile workforce can locate installed assets with centimeter accuracy, meaning individual meters formerly sharing a parcel can now be uniquely identified.

The new GIS-based workflows are user-friendly, enabling the mobile workforce to expand their project scope with the capability to map new hydrant and valve installations in addition to the city's water meters. Faster response times have improved customer service for the residents of Santa Barbara.

"Drought conditions certainly challenge municipalities like ours," Lancy said. "Efficiently maintaining our system is vital. Leaks need to be located quickly and addressed promptly. Having the right tools to identify, locate, and document repairs means better customer service and reliability—which is very important to us."

This story originally appeared as "Santa Barbara Strengthens Field Data Collection with a New Mobile GIS Workflow" 2020 on esri.com. All images courtesy of the City of Santa Barbara unless otherwise noted.

GETTING STARTED WITH GIS

GOVERNMENT OPERATIONS BECOME MORE EFFICIENT WHEN decision-making includes location information and spatial data-driven tools. Location intelligence derived from visualization and analysis of geospatial data opens additional paths of understanding and new insights about the distribution of data across geographies. For example, by adding GIS layers of data to decision processes, such as demographics, traffic, and weather, state and local governments start to see why things happen where they do and how historical patterns can be used to prepare for and even predict problems. Perhaps the most powerful aspect of using GIS maps and analysis for decision-making is the rapid increase in operational efficiency. GIS maps and apps put operational managers and mobile workers and technicians on the same playing field, allowing them to work more effectively together, while reducing the time to complete tasks and avoiding unnecessary costs.

To get started using GIS for improving operational efficiency, state and local governments typically begin with a practical approach to problem-solving, using one or more of the following methods.

Coordinate mobile data collection with GIS apps

GIS apps on smartphones and devices allow you to not only document the location of assets and observations but also collect important details, such as photos, videos, and audio recordings, all of which can be included in interactive maps. Mobile workers, with little or no GIS experience, can use mobile GIS data collection apps to generate asset inventories, perform on-location inspections to update records and status, and alert managers and other mobile staff about problems and concerns about physical and environmental conditions. Improving data collection with GIS ensures that asset management and operations systems are up to date and that the mobile workforce is working with the latest information.

Use GIS dashboards to monitor performance and track progress

GIS operation dashboards are interactive web applications that present a variety of data measurements on a single screen. GIS dashboards are used within operations centers to monitor project or work status over time, report current physical and environmental conditions of assets, and alert managers and other operational staff about potential threats and risks to assets, mobile workers, and the public at large. Dashboards are often used to inform community leaders and the public about current conditions for issues, such as road work and closures, status of ongoing projects, and government response to and recovery from storms, flooding, and other severe weather-related events. Dashboards are also employed to monitor impacts from disease and other health-related issues, such as COVID-19 and opioid addiction. Operations managers use GIS dashboards to track key performance indicators (KPIs) and make decisions by evaluating location-based data against strategic initiatives and efficiency goals and objectives.

Deploy a GIS hub to engage residents

A GIS hub is a website that is connected to location-based data, maps, online forms, and citizen reporting tools. The purpose of a GIS hub is to help residents understand the scope and impact of government initiatives and how those efforts affect their lives, property, and businesses. GIS hubs often provide a steady stream of public input through location-based surveys and mobile problem reporting apps, which increase a government department's ability to base operations on the needs and desires of the community and create opportunities for citizen participation and public feedback on operational performance.

Using GIS

There are two ways to get started using GIS: hands-on learning and applying ArcGIS Solutions.

Hands-on learning

Hands-on learning will strengthen operations managers' and staffs' understanding of GIS and how it can be used to improve operational efficiency. A good place to start is with Learn ArcGIS, an online resource that introduces GIS using practical scenarios. Learn ArcGIS lessons demonstrate how location-based operational tools work when applied to real-world problems.

- Try a selection of ArcGIS mobile apps, such as ArcGIS Collector and ArcGIS QuickCapture. Learn how GIS mobile data collection apps work, how to create maps used in mobile apps, and how to generate maps and dashboards from data gathered with mobile apps.

- Get a feel for managing successive stages of a wide-scale mobile operation, including drive-by reconnaissance and on-location inspection of assets. See how to edit and update asset status in the field. Analyze the density of collected locations to prioritize asset replacement work.

- Create and deploy your first GIS dashboard. Learn how to customize the dashboard elements to work with real-time data, add charts and maps, and add meaningful information through indicators and metrics that communicate effectively with your audience.

- Learn how to get started with the location-based Citizen Problem Reporter solution, used by public works departments, to solicit reports from residents, manage the response to each report, get feedback from the public, and monitor the resolution of nonemergency problems.
- Get started with drones to capture high-resolution, location-based imagery of operational events and project areas. Learn how to convert drone data to map information and create 3D representations of buildings.

Learn about additional GIS resources related to state and local government operational efficiency by visiting the web page for this book at **go.esri.com/bsc-resources.**

ArcGIS Solutions

ArcGIS Solutions is a collection of focused maps and apps that help address operational challenges in your organization. As part of the Esri Geospatial Cloud, solutions work with your data and are designed to improve operations and gain new insights. Common solutions for government operations give you the ability to do the following:

Monitor key performance metrics and communicate progress

The Performance Management solution delivers a set of capabilities that help you tabulate performance metrics, monitor performance using a series of dashboards and key performance indicators (KPIs), visualize performance for defined geographies, and share performance with the public through a community website.

Streamline road data management and improve road data quality

The Road Network Data Management solution delivers a set of capabilities that help you maintain a standard set of public road layers, streamline road data management workflows, track maintenance agreements, and continuously improve the quality of public road information.

Create right-of-way asset inventories

The Right of Way Asset Inventory solution is a set of GIS applications for creating and managing visible government assets, such as signs, traffic signals, streetlights, bridges, railroad crossings, sidewalks, bike lanes and paths, and parking.

Track and communicate capital project status

The Capital Project Tracking solution is a set of capabilities that help you manage the status of active capital projects, track project performance, share project progress with internal stakeholders, and communicate investments being made with the public.

Proactively manage and share road closures and detours

The Road Closures solution delivers a set of capabilities that help you inventory road closures and related detours, communicate closures and detours with the public, and share authoritative closure information with consumer applications, such as Waze.

Respond quickly to mosquito service requests

The Mosquito Service Requests solution delivers a set of capabilities that help you solicit mosquito activity reports from the public, triage reports to staff and field crews for resolution, and monitor the resolution of reports in a community.

Empower residents to report problems

The Citizen Problem Reporter solution is a set of capabilities that help you solicit problem reports from the public (such as graffiti, trash, potholes, and blight), manage the response to each report, receive feedback from the public after addressing a report, and monitor the resolution of nonemergency reports in a community.

Takeaways

With GIS, state and local governments can transition operational processes away from inefficient methods and legacy systems. Governments can use GIS data collection tools to improve the accuracy and relevance of asset inventories and inspections. GIS provides government departments—such as public works and engineering—the ability to increase response effectiveness and monitor the performance of projects and initiatives.

In the case studies section, you learned how the District of Columbia Department of Transportation created and used a GIS-based performance management tool to monitor and provide status reports of pavement work and the work its staff accomplished each week. The City of Salinas, California, used GIS to radically change how stormwater is managed, giving the city's public works department the ability to collect smarter data and streamline the workflows that reinforce its regulatory reporting requirements. The maintenance staff at Ontwa Township in Michigan used GIS mobile apps to assign work orders, provide updates on completed tasks, and access service orders and preventive maintenance plans. They shared the information with managers and township officials, strengthening Ontwa's commitment to transparency. The City of Santa Barbara, California, replaced an out-of-date legacy system with GIS applications to meet its immediate need for an improved, more accurate mobile data collection workflow.

Learn ArcGIS is a way to quickly learn about GIS tools, such as mobile data collection apps, dashboards, and citizen problem reporter solutions through real-life scenarios. ArcGIS Solutions provides operational departments at state and local government levels opportunities to start using GIS to monitor key performance metrics, streamline data management, improve data quality, create and update asset inventories, share information about road closures and detours, respond faster to citizen requests, and empower residents to report nonemergency problems.

Next, you will learn about using GIS to make data-driven decisions.

PART 3

DATA-DRIVEN PERFORMANCE

STATE AND LOCAL GOVERNMENTS USE GIS TO UPDATE workflows with new technology, including artificial intelligence, machine learning, and sensors. GIS expands a government agency's ability to determine who is impacted most by issues—such as at-risk populations during the coronavirus disease 2019 (COVID-19) pandemic—and decide where government assistance will have the most beneficial effect. Today, government agencies apply a location data approach to drive a wide range of activities, such as the following:

- Tracking hot spots of disease outbreaks
- Monitoring operations outside of the office
- Managing budgets and capital improvement projects
- Understanding social issues and equity across neighborhoods
- Identifying areas for environmental and economic restoration

Ultimately, GIS maps and analyses are used for decision-making. Good decisions are based on accurate and timely information, which means that decision-makers need ways to measure and assess performance so that they can determine whether goals and objectives are being met on time, within budget, and across the communities they serve. Based on relevant information and high-quality data, decision-makers need to see where strategies and programs should be adjusted, where resources and activities should be redistributed, where progress is being made, and where efforts are falling short.

In the context of data-driven performance, GIS enables governments to collect and organize vast amounts and types of data, analyze it quickly, and empower departments and agencies to move from a nongeographical viewpoint to real-time, location-based decision-making.

Case studies

Data-driven decision-making is a process that state and local governments use to validate and prioritize projects and initiatives. At issue is the sheer volume of data that floods into government agencies on a daily basis and tools that analysts and officials use to turn raw data into meaningful information. Governments are increasingly turning to data visualization solutions, such as GIS, to accelerate decision-making and build trust with their constituency.

In the following case studies, you will see how governments and agencies have elevated their approach to data management and analysis by adopting a more location-centric strategy using GIS. Through location intelligence, they have discovered an increased value in data and improved their ability to ensure confidence and reliability in the decisions they make.

CREATING A SINGLE, SHARED SOURCE OF INFORMATION

Texas Parks and Wildlife Department

REGIONAL DIRECTORS AND PARK MANAGERS IN THE TEXAS Parks and Wildlife Department (TPWD) are long-time proponents of using data to guide their decisions. However, prior to using GIS, data analysis and reporting was hampered by manual processes and the lack of a single, shared source of information.

Previously, TPWD would send a monthly email to stakeholders, which included a spreadsheet that reported the costs and revenues for state parks and campgrounds showing statistics for the most recent month and compared data to the same month in the previous year. While the email and spreadsheet provided a valuable snapshot of park and campground performance, additional analysis required copying and pasting data from historical spreadsheets, a time-consuming and inefficient method that didn't lend itself to timely and effective decision-making.

For example, a park manager trying to decide the best allocation of limited maintenance funds would want to know which camping facilities served the most visitors or generated the most revenue over a specified period of time. In such a case, a park manager would review old spreadsheets and make rough estimates for future performance. Typically, a park manager worked with their own spreadsheet copies and performed their own analysis, often using different types of datasets and methods of analysis.

Today, the Texas State Park Insights Viewer, a GIS-based application, now serves as a single source of information for all park managers reporting on revenue, visitation, and capital construction data. The Insights Viewer is an intuitive GIS dashboard that enables regional directors, park managers, and other employees to conduct

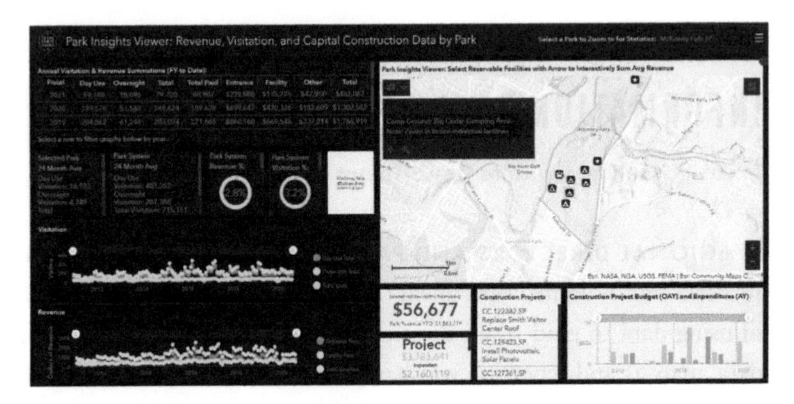

The Texas State Parks Insights Viewer enables staff to evaluate park performance.

the kind of detailed analysis needed to evaluate performance and make data-driven decisions.

With the Insights Viewer, TPWD can calculate how much revenue would be lost if certain campgrounds or camping loops were shut down for construction projects. Such analysis requires running multiple scenarios on each construction project to determine where and when the projects should be scheduled to minimize the loss of revenue. Seeing the location of projects on a map, along with costs and revenue projections, helps TPWD understand the potential impact of construction projects locally and across the entire state parks and campgrounds system.

"Bottom line, the useful information provided in Park Insights Viewer allows us to make informed decisions quicker than ever before," said Reagan Faught, State Parks Region 2 director for TPWD.

This story originally appeared as "GIS Dashboard Reengineers Planning at Texas State Parks" in *How Data-Driven Performance Powers Smart Communities Industry Perspective*, Govloop, 2021. All images courtesy of Govloop and Esri unless otherwise noted.

IMPROVING FINANCIAL TRANSPARENCY

City of Topeka, Kansas, Public Works Department

THE PUBLIC WORKS DEPARTMENT FOR THE CITY OF TOPEKA, Kansas, designs, builds, renovates, and operates public projects. These range from bridges, traffic signs and signals, and city-owned trees to streets, sidewalks, and parking meters and garages. The department strives to deliver programs and projects that enhance public health and the quality of life for everyone who lives in and visits Topeka.

Because the department relies on public funds, it has to collect and share accurate data about its programs and projects to enable leaders to make informed decisions about how funds are spent and increase transparency with the public. Tracking this information is the responsibility of the department's technical support group (TSG) division.

Toward the end of 2019, the TSG upgraded how it displays the Public Works Department's budget information so the public could understand where funds are allocated and others at the city could more easily see how finances are distributed among various departments. The TSG chose to use ArcGIS Insights for its advanced analytics capabilities and engaging visual displays.

According to TSG Division Manager Brandon Bayless, various divisions and departments within the City of Topeka were not sharing financial information, which meant there were likely no larger discussions going on about the data. This made it difficult for different units to collaborate and get the details they needed to make data-informed decisions. What's more, residents didn't have a good idea of how public funds were being used.

"For the city, there's been a big push to have reliable, timely, and accurate information [to help] make better decisions based on

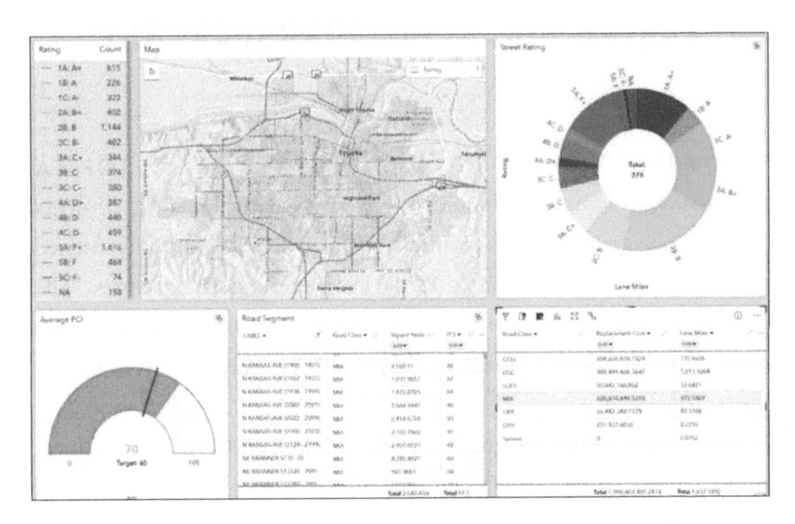

The City of Topeka's asset management program uses ArcGIS Insights to track the condition of all city assets and join that information to other data in its enterprise GIS platform.

support data," said Bayless. "A lot of the information was not necessarily being shared or analyzed." The data was there, he said, but were various departments able to infer anything from it?

Bayless explained that ArcGIS Insights was chosen as the solution to this problem because it had all the tools the TSG needed to display data in a web-based platform. Darren Haag, a solutions architect in the TSG division, agreed. "When I was going through and building different pieces for open data, I knew I wanted to utilize ArcGIS Insights, and I thought the budgets were a good [way] to use it," said Haag. "Even though we don't necessarily need to map that data, I felt like ArcGIS Insights is a powerful business intelligence-type application that we could use for financial data." With ArcGIS Insights, the TSG can bring in information that, when applied to real-world situations, gives the city a more holistic picture of what's going on.

"Everything that's done around the city, whether it be fixing potholes or fixing water main breaks, has a location," said Bayless. "Being able to analyze that and show people on a map or [provide] information [visually] is extremely beneficial." With ArcGIS Insights, the TSG division can provide city employees and the public with more than just numbers on a spreadsheet. It can present information visually, making it easier to understand.

Within the Public Works Department, ArcGIS Insights has improved data collection and analysis. According to Bayless, managers used to spend the majority of their day trying to access and extract data using the city's financial software, which produced spreadsheets and standard text reports. It took them from an hour to an hour and a half to pull the data, create a .zip file, and post it to an FTP site. Now, however, managers can easily collect any data requested by the public, city council members, or others and make it available to download via a link on the city's open data website.

"ArcGIS Insights reduces the time needed to pull data. Now it's a very simple email response," said Bayless. "That's one of those efficiencies that makes things work smoother because we're not having to put pressure on somebody to do all this manual work. In the short time we have been using ArcGIS Insights, it has become apparent how much more efficiently our time will be used in exploring and analyzing our data."

The Public Works Department uses ArcGIS Insights to house the budgets and financial data for three of its divisions. This has saved time and improved accuracy. The ease of data input and analysis has enabled division managers to better track their actual spending, analyze trends, and find areas where additional funding is needed or where budgets can be reduced. "The data and analytics that we provide with ArcGIS Insights or that we empower other departments to

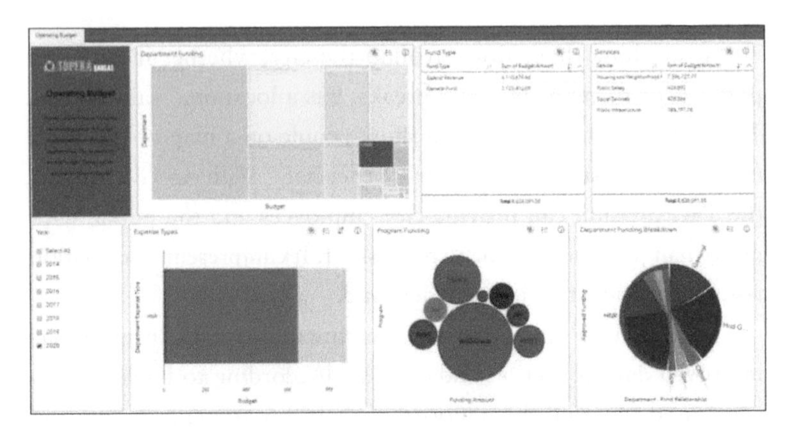

Having budget information available in Insights allows division managers to better track spending and analyze trends.

use help them make better decisions, which benefits the community," said Bayless.

For residents, having access to the eye-catching charts, graphs, and tables in ArcGIS Insights empowers them to find answers to their questions. This frees up city staff so they can focus on the services they provide. Haag said ArcGIS Insights is easier to use and better organized than other tools the department tried. While he believes that it could elicit more questions from residents once they review the data, Haag welcomes those questions because they help staff know what's important to people.

"They possibly see trends or look at things and ask questions like, Why is this showing this? or Why aren't you doing something here?" said Haag. "It opens up that line of communication, which can help make the organization work better."

This story originally appeared as "Data Visualization App Increases Financial Transparency" in the Summer 2020 issue of *ArcNews*. All images courtesy of the City of Topeka unless otherwise noted.

INCREASING DATA VALUE WITH LOCATION ANALYTICS

City of Sacramento, California

FOR MANY AMERICANS, INCLUDING CALIFORNIANS, THE city of Sacramento is primarily the location of California's statehouse. For Sacramento's residents, it is a thriving city in its own right—by some metrics, the fourth most ethnically diverse city in the United States.

What the city's varied communities have in common is a high engagement with Sacramento's 311 service. In any year, the service logs around 500,000 different interactions—or one for every Sacramento resident.

Since Sacramento launched the 311 service in 2008, the city has maintained an ambitious conception of what the program should achieve. While most large- or medium-sized cities maintain a 311 program, many of them are structured as adjunct city agencies that perform a kind of ombudsman role. In Sacramento, however, 311 is conceived of as a civic connective tissue, a digital hub that links many city departments.

"It's the front door to Sacramento," said Maria MacGunigal, the city's chief information officer (CIO), "the highest touch point for all the interactions the city has with the community."

In 2013, soon after MacGunigal accepted the job of Sacramento's CIO, 311 was placed under her purview.

"That isn't typical for a city's IT department, but the 311 system was struggling under heavy demand," she said. Much of the challenge involved the ambitious scope of Sacramento's 311 program, which for many years has used GIS to route requests.

"We've always had a back end ArcGIS Server that supported geocoding and some overlay values," said Dara O'Beirne, the city's GIS developer. "Whenever someone submits a ticket and enters an address into the interface, it validates against our internal geocoder to ensure that it actually is a valid address within the city of Sacramento. Then it conducts an overlay and pulls attribute information from 35 different layers, such as the correct council district or police beat."

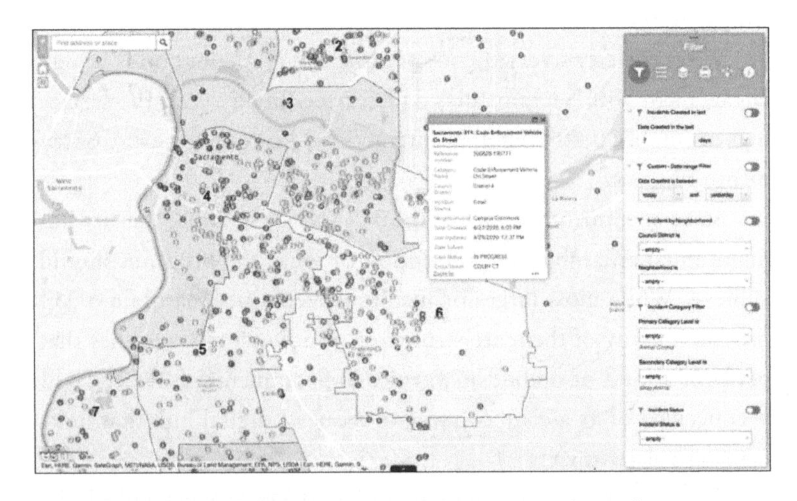

Users of Sacramento's new 311 system can now see incident tickets on a public-facing map.

A report of a stray dog, for example, would automatically note the relevant animal care district and notify animal care officers in the area. The ticket also generates automatic updates on the problem for the public and any agencies involved.

"Right now, that's occurring from Salesforce, through our firewall, and into ArcGIS Server," O'Beirne explained.

"Most cities don't have that kind of back end integration through all those business lines," MacGunigal added. "It doesn't have nearly as much of an impact on the community if we're just taking notes and handing off the requests."

When MacGunigal was tasked with overhauling Sacramento's 311, she wanted to broaden the program's scope even further so it would eventually become "a foundation upon which we can build all other portal access to the public," she said. That meant retaining the system's GIS-enabled back end while also using GIS to improve the real-time interface. With the new setup, 311 users interact with maps, via Salesforce, that are built using ArcGIS API for JavaScript.

Consider that stray dog. The first person to report it will see the ticket displayed on the map. If someone else sees the dog and logs on to 311 a few seconds later, that person will see the same ticket and realize there's no need to report it.

"Managing multiple reports [of the same issue or incident] was one of the challenges we were trying to overcome with this integration between Salesforce and ArcGIS Online," O'Beirne said. "As soon as someone submits the ticket, it goes to ArcGIS Online, so it'll be reflected on the map. The next person can see the same incident at the same location. If they click on it, they open the ticket in Salesforce to find out more information." People can even follow the ticket to receive notifications when updated information is added to the ticket. "It's a more encompassing experience for the user," O'Beirne said.

The system also helps city agencies better serve their constituents. "We now have data from GIS maps and layers, and we take the Salesforce data and start building dashboards around it," said Ivan Castellanos, Sacramento's 311 manager. "There's even more value in the data now because it can be used to help various city departments drive their strategies and make data-driven decisions."

"From my perspective, one of the most important aspects of this implementation is not that any one of the components of the GIS integration is brand new, but that they're as comprehensive as they are across all the different levels of integration with the map or with the geography," MacGunigal said. "For example, the idea that you would try to geographically understand if something had already existed or it had already been reported has been around for a long time, but we just used to do it mathematically. We didn't actually use GIS. We used an approximation of what might be in proximity, but it became so burdensome in the system that we actually removed it at one point. So, the concept was there, but the full implementation wasn't quite there until now."

"I think at the beginning of the project it was difficult because it was a completely unknown endeavor," O'Beirne said. "There weren't that many people out there who had already done this. There were one or two other cities, but not to the level that we were trying to implement."

The final challenge was unexpected. When California's governor, Gavin Newsom, issued the state's first stay-at-home order in response to the COVID-19 crisis, MacGunigal's team was nearing the end of several months of planning for the 311 transition. After all the work they'd put in, was the effort to push the project over the finish line while the team was scattered worth jeopardizing it? MacGunigal chose to power through. The new 311 debuted on April 15, a mere month behind schedule. All feedback suggested that any hiccups, inevitable in this kind of launch, were minor.

The new technology worked so well that the team was able to surmount an unforeseen hurdle. Operators trained to answer calls in the 311 call center were among city employees who had been sent home. In addition to launching the new system remotely, the team also had to figure out how to make the 311 call center function remotely, a daunting task made simpler by the cloud-based architecture of Salesforce and ArcGIS Online.

"It's not unheard of in the private sector to have call centers that have a lot of distributed locations, whether that's at people's homes or just multiple locations for call taking, but we're one of the first public agencies to have a mostly remote call center," MacGunigal said. "Everyone was remote—the development team, the GIS team, the infrastructure team—and we pulled it off without a hitch. So that was pretty awesome."

This story originally appeared as "Sacramento's Revamped 311 System Uses Groundbreaking Location Analytics" in the Fall 2020 issue of *ArcNews*. All images courtesy of the City of Sacramento unless otherwise noted.

MEETING THE NEEDS OF MODERN RESIDENTS

City of Coral Gables, Florida

CORAL GABLES, FLORIDA, SITUATED ALONG MIAMI-DADE County's bayfront coastline, was one of the first planned cities in the United States. Established in 1925, it was inspired by the City Beautiful movement of the late-19th and early 20th centuries—an urban design concept that maintained that a community's design couldn't be divorced from its social issues.

In continuing to heed this vision, the tree-dotted, Mediterranean-style community is applying smart city concepts to improve the services it offers to residents, businesses, and visitors. To achieve that, Coral Gables recently launched an open data site on ArcGIS Hub that it uses to provide convenient access to city data and services, advance important initiatives, and improve quality of life for residents and tourists.

Coral Gables has been long acclaimed for its beauty and livability. Given its stature as a leading planned community, the city is embracing technology and innovation with the aim of becoming a truly smart city.

For the past three years, Coral Gables has been working to align its road map and strategic plan with its mission: being a world-class city with a hometown feel. To that end, the city's IT department has been busy building a digital ecosystem of people, businesses, organizations, systems, and other things that contain, promote, and sustain that smart city culture.

In early 2018, IT staff used ArcGIS Hub to launch an open data site for Coral Gables. Now, anyone can view the city's financial, legislative, permit, and other public records via a variety of new and

On the Coral Gables Smart City Hub, anyone can view the city's financial, legislative, permit, and other public records via a variety of new and existing services.

existing services. Introduced in April, the Coral Gables Smart City Hub receives more than 300 visits per day.

The hub's purpose is to simplify how people access city services and information that would typically require a call or visit to one of the city's offices. It is just one of several interconnected and interoperable elements—including other data platforms, the Internet of Things (IoT), and robust high-speed communication networks—that is beginning to transform how the city and its residents communicate with each other.

The glue that binds all this together is GIScience, which enables Coral Gables' IT department to horizontally integrate apps and data—removing data from silos, finding and filling data gaps, and resolving redundant and duplicate data. Additionally, IT staff employ hybrid clouds and hyperconverged infrastructure to enhance the hub's capabilities for artificial intelligence (AI) and machine learning.

An advantage of having ArcGIS Hub is that the City of Coral Gables can put a variety of community engagement initiatives in the platform. One of the city's first ArcGIS Hub initiatives is a collaboration with the University of Miami's School of Architecture and Center for Computational Science that invites participants to design their own smart city solutions for Coral Gables. The initiative provides contestants with access to datasets and a platform that teams can use to collaborate on their designs. Participants are asked to create prototypes of technology solutions and apps for known transportation and traffic issues with the aim of improving residents' quality of life and visitors' experiences.

"Bringing more convenience and better quality of service [to] our citizens, that's the main driver" behind the hub, said Raimundo Rodulfo, the IT director for Coral Gables. "It's quality centric."

That's something that a lot of communities, not just Coral Gables, want to provide for their residents. Implementing ArcGIS Hub helps residents communicate their concerns with city stakeholders and collaborate more effectively with them to come up with solutions. In turn, this makes it easier for cities to implement technology that can make discernible improvements in safety, transportation, convenience, and government processes.

As with most cities, ensuring public safety is paramount for Coral Gables. With ArcGIS Hub in full use, the city now has an array of collaborative policing and emergency preparedness tools that it didn't have before.

Smart policing is one public safety initiative that has benefited tremendously from using hub technology. Residents can access police data through the Smart City Hub, and the GIS layers available via

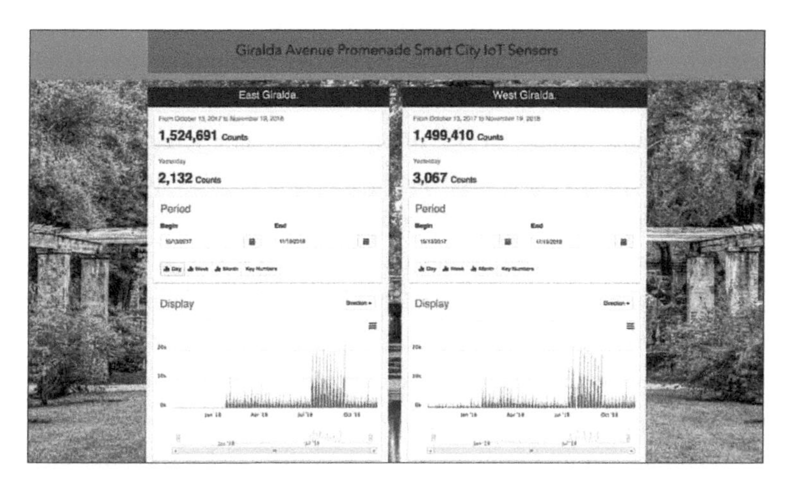

Data platforms, robust high-speed communication networks, and - devices powered by the Internet of Things—such as the pedestrian sensors deployed on the promenade in Giralda Plaza—are transforming how the City of Coral Gables and residents communicate with one another.

the hub can be referenced by or integrated with Coral Gables' smart policing systems.

The Crime Intelligence Center (CIC), for example, uses a stack of GIS-based and GIS-reliant technology to help reduce and prosecute crime within and near Coral Gables' boundaries. Built by the city's IT department, along with the police department and other public safety teams, the CIC employs closed-circuit television (CCTV) to find and track suspects within the city; a video analytics tool, called BriefCam, that compresses all the video data and uses AI and machine learning to find patterns; and CrimeView, an app based on ArcGIS, to map and analyze crimes. Top officials from all city departments engage with the data from these sources in biweekly strategy sessions, where they discuss how to respond to and plan for various types of crime. The use of these technologies in concert with one another has contributed to a 30 percent decrease in crime in Coral Gables over the last two years. And residents have access to more information about all this via the Smart City Hub.

Hurricane preparedness is another public safety issue that has benefited from having hub technology. When tropical storms come through Coral Gables, city staff, residents, and visitors work together to gauge the threat, commit adequate resources to it, assist in safeguarding lives (by conducting evacuations and identifying shelters), and ready robust responses. ArcGIS Hub makes it easier for people to gain access to vital mapping and data tools so they can collaborate in managing these environmental events.

Planning for and managing an innovative and sustainable city require stakeholders from various departments to agree on targeted actions, adhere to schedules, properly allocate resources, and acquire adequate funds. For Coral Gables, its Smart City Hub is becoming an integral part of that process. For example, the city embarked on an ambitious sustainability plan to minimize energy expenses and gain energy efficiency. Some of the objectives of this plan include converting the city's public service fleet (for example, sanitation vehicles) to electric to reduce gasoline consumption, making water-saving upgrades to green space and landscape irrigation systems, and advancing Leadership in Energy and Environmental Design (LEED) building practices. In three years, this plan has had some concrete successes. For example, under the Property Assessed Clean Energy (PACE) program, which helps homeowners and business owners finance renewable energy and energy efficiency upgrades, Coral Gables has completed more than 200 energy mitigation and building improvement projects at a value of nearly $7 million.

The City of Coral Gables can report metrics and impacts of the newfound efficiencies dynamically via the Smart City Hub. All the information is available on dashboards, which local residents, business owners, and even visitors can use to see where these improvements have taken place throughout the city and how they have affected critical resources such as air and water quality.

"All that [statistical] data becomes actionable information—for business development and commercial purposes but also for regular citizens and those who know metrics and data," said Rodulfo.

This story originally appeared as "Smart City Hub Meets the Needs of Modern Citizens" in the Winter 2019 issue of *ArcNews*. All images courtesy of the City of Coral Gables unless otherwise noted.

GETTING STARTED WITH GIS

DATA-DRIVEN DECISION-MAKING BEGINS WITH exploration and analysis of data. For most organizations, the goal is to gain a deeper understanding of what is going on, identify problems, and devise solutions that address the most severe issues. State and local governments have the added responsibilities of providing residents with a transparent view of the decision process and helping them understand why an issue is important, not only to them as individuals but to the community as a whole.

With GIS maps, spatial analysis, and applications, residents and government leaders find a common ground for decision-making. Instead of relying completely on numbers and statistics, discussions around projects and initiatives become centered on where actions are taking place and who or what will be affected by that action. The interactions among residents, business leaders, institutions, and government leaders transform into a shared dialogue.

Getting started with data-driven decisions using GIS means reframing a few basic questions.

How will you create and share authoritative data?

State and local governments are composed of multiple departments, often distributed across many locations, that work on the same projects and initiatives. Unfortunately, dispersed departments tend to use their own datasets and methods to analyze data and make or contribute to decisions. GIS serves as a unifying support system that organizes data and analysis methods by location, providing access to the same authoritative data, including maps and geostatistical tools. Decision-makers can compare budgets, revenue, service areas, visitor information, and other measurements across local and regional geographies, combining standardized and spatial methods to provide sensible real-world results.

How will you improve data accuracy and data collection methods?

A decision is only as good as the data it is based on. Poor data accuracy and unreliable data collection methods undermine data-driven decisions and erode trust with internal stakeholders and the public. GIS can record and maintain data down to survey-grade accuracy and allow that information to be aggregated into larger geographies, such city, county, and state boundaries. With GIS data collection applications, mobile workers can use smart devices and survey instruments to capture precise locations and complete standardized forms and surveys that feed directly into government databases. GIS can also be used to combine streaming data from devices, such as smart sensors, with the advanced analysis and processing capabilities of machine learning, which allows government agencies to rapidly detect, prioritize, and respond to requests and incidents.

How will you communicate data-driven decisions to those who need to know?

When a government official presents a project or initiative to the public, the decision to move forward involves the inherent risk that it may be misinterpreted, causing delays. A large part of the problem is that residents want to know what this decision means to them. Others may feel that they have not been kept informed leading up to the decision. With GIS maps and dashboards, data analysis and representation can be broken down into meaningful chunks that show where the issues are and how the information is relevant to the audience. Residents, business owners, and other stakeholders can see where advances have taken place in the community and what resources are available to them. As an example, public-facing websites built on ArcGIS Hub provide an interactive experience that guides visitors through the decision process, presents the value of community projects and initiatives, and provides opportunities for community feedback.

Using GIS

There are two ways to get started with GIS: hands-on learning and using ArcGIS Solutions.

Hands-on learning

Hands-on learning will strengthen your understanding of GIS and how it can be used to improve data-driven decision-making. A good place to start is with Learn ArcGIS, an online resource that introduces GIS using real problems and scenarios. Learn ArcGIS lessons will help you understand how data-driven decisions are represented and learn more about the following:

- Creating an ArcGIS Hub site to share open data and other content (such as datasets, apps, surveys, documents, and web maps), find patterns in civic data, and encourage citizen participation
- Using ArcGIS Insights to analyze complex datasets and create clean, helpful infographics for reporting results, including charts and maps
- Constructing location-enabled surveys with ArcGIS Survey123 that you can use in reports, websites, and ArcGIS Hub
- Monitoring real-time data with ArcGIS Dashboards to track mobile data collection, streaming data feeds from sensors, and visualizing citizen requests

- Evaluating readiness by converting objectives into metrics that can be used in data-driven analysis for assessing and monitoring impacts and effectiveness

Learn about additional GIS resources related to state and local government data-driven performance by visiting the web page for this book at **go.esri.com/bsc-resources.**

ArcGIS Solutions

ArcGIS Solutions is a collection of focused maps and apps that help address data challenges in your organization. As part of the Esri Geospatial Cloud, solutions work with your data and are designed to improve data quality, increase confidence in decisions, and share results with internal stakeholders and the public. For example, you can do the following:

Aggregate authoritative community data

The *Community Data Aggregation* solution helps transform and load source data, such as addresses and parcels data, into a GIS information model that helps state and local governments organize authoritative data for emergency response, permitting, land use, and regulatory issues.

Curtail declining property values in neighborhoods

The Neighborhood Stabilization solution helps state and local governments conduct property surveys, measure the fragility of neighborhoods, track blight, see demolition activity, and publish a focused set of information products, including a Tax Liability Calculator, a Property Condition Survey, and a Neighborhood Early Warning analysis tool.

Increase transparency and crowdsource public feedback

The *Participatory Budgeting* solution is a public-facing application that helps drive community engagement by crowdsourcing capital improvement ideas from residents and then routing that information to the right government staff or responsible department. Residents can use the Participatory Budgeting Hub site to learn about the budgeting process, as well as identify, discuss, and prioritize project ideas.

Takeaways

GIS enables governments to collect and organize vast amounts and types of data, analyze it quickly, and empower departments and agencies to move from a nongeographical viewpoint to real-time, location-based decision-making.

In the case studies section, you learned how the Texas Parks and Wildlife Department built upon their affinity for using data to guide their decisions by creating a GIS-based application that now serves as a single source of information for all park managers reporting on revenue, visitation, and capital construction data. The City of Topeka employed GIS as a business intelligence application for understanding financial data in real-world situations, providing a more holistic picture of what's going on. The City of Sacramento completely revamped their 311 system by combining GIS with Salesforce, making it possible for their operators to continue working from home during the COVID-19 pandemic. The City of Coral Gables used GIS to create an array of collaborative policing and emergency preparedness tools and share metrics and improved efficiencies with residents and business owners.

Learn ArcGIS is a way to quickly learn about GIS tools for data-driven decision-making, including sharing open data, analyzing complex datasets, creating location-based surveys, and monitoring real-time data feeds. ArcGIS Solutions provides state and local governments with a variety of rapidly deployable tools for decision-makers, including building authoritative community data and crowd-sourcing capital improvement ideas from residents.

Next, you will learn about using GIS for civic inclusion.

PART 4

CIVIC INCLUSION

COMMUNITY RESIDENTS CARE ABOUT WHERE THEY WORK and live. They want access to quality schools, safe neighborhoods, clean drinking water, and the assurance that government leaders hear and respond to their concerns. To respond appropriately, state and local government agencies must fully understand the people they serve, so that they can provide opportunities for all without leaving anyone behind. GIS is an instrumental part of building a community that emphasizes civic inclusion.

Maps and spatial analysis promote a location-centric point of view that is critical to meeting the needs of residents. State and local governments use GIS to engage and collaborate with residents in creative and informative ways, but they also express the value of what's being done—through data—to support all residents.

As with any data-driven decision-making, civic inclusion should be driven by goals that can be measured and revised as needed. At its core, civic inclusion is about providing transparency, keeping the public informed, encouraging public participation, and placing an emphasis on equity and social justice. By including the public early and often in decision-making, community leaders provide greater accountability and transparency.

The following case studies are only a few examples of how state and local governments are using GIS to develop a deeper understanding of their communities and actively pursue the goal of total civic inclusion.

Case studies

GIS provides a different way for governments to think about and practice civic inclusion. With GIS, state and local government leaders and professionals have an extra tool for pinpointing locations where people's voices might be overlooked. It can help raise questions such as where are people speaking up and where are they not? Which neighborhoods are at risk of falling behind? How can a connected citizen act as a sensor to help governments keep on the right course? Mapping and spatial analytics help governments better understand their community makeup and help residents understand what happens and why, in a context of where they live.

The following case studies show how state and local governments are using GIS to develop a better, more transparent relationship with residents by creating open data portals and initiative-driven online hubs and getting their residents involved in decision-making.

REACHING MORE PEOPLE WITH ACCESSIBLE DATA

City of Johns Creek, Georgia

IF YOU WANT TO KNOW WHEN THE NEXT CITY COUNCIL meeting will be held, you live in Johns Creek, and you have an Amazon Echo, just ask Alexa, Amazon's voice control system for its wireless Echo smart speakers. Alexa can respond appropriately to your question because the City of Johns Creek, Georgia, developed an Alexa skill that makes key data about the city's operations more accessible to residents via natural language questions. An Alexa skill is like an app that performs a specific task, such as using voice commands to perform everyday tasks, like checking the news. Johns Creek, the first city in the world to use its open data with Alexa, won the Best Practices Award for Amazon's City on a Cloud Innovation Challenge in June 2018. The development of the Alexa skill is just the latest sign of the city's commitment to using GIS and open data to enhance city services and drive its economic development.

Johns Creek has an exceptionally large entrepreneurial population—half the city's licensed businesses are home based. From its inception, Johns Creek has relied on technology, including GIS, to deliver a high level of service to residents.

Nick O'Day, chief data officer, believes that the city realizes value from the democratization of information that the city's open data site makes possible. Reaping the benefits from open data involves first making everyone—the public, businesses, and developers—aware that the city has high-quality data that is freely available.

Initially, the city created the Johns Creek Open Data Portal using ArcGIS Open Data to give the public free access to high-quality GIS data. The open data portal provides better government transparency and also enables businesses from mom-and-pop entrepreneurs to large corporations to use GIS data. Access to data that is trustworthy and free mitigates many of the risks faced by startups and small businesses.

Johns Creek expanded their open data strategy by launching the DataHub, which incorporates nonspatial as well as spatial data and serves as a one-stop destination for all city data. The DataHub also provides interactive tools that help users discover, understand, and work with large amounts of city-generated data, including, videos, dashboards, charts, web maps, and smart mapping and spatial analysis tools. For example, the DataHub offers videos that show how to use the data to locate potential customers, find vacant properties and commercial sites, discover restaurants along a commute, and incorporate data services into an app.

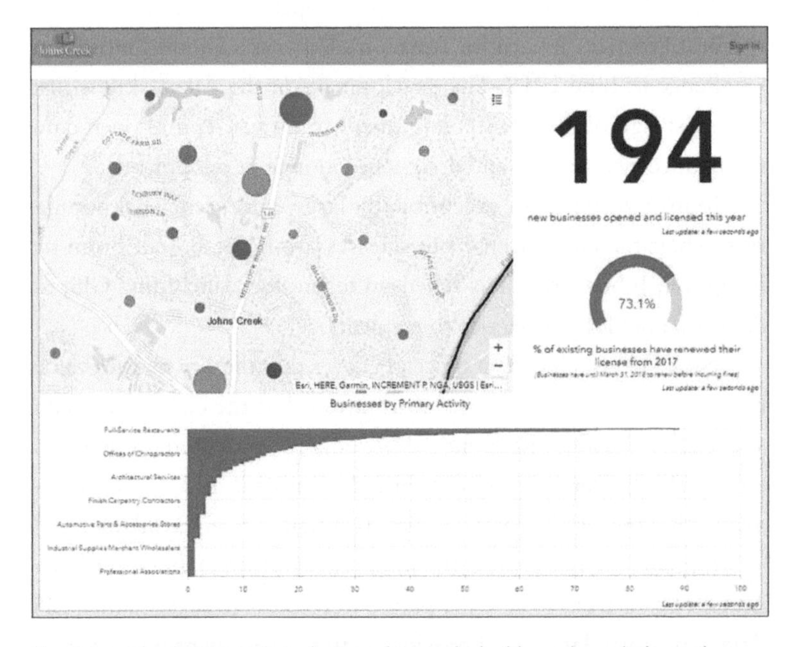

The Johns Creek DataHub is designed to push dashboards and charts that make data easy to understand.

O'Day says the site has been engineered specifically to push dashboards and charts first and data second because they "want to drive understanding through the data, instead of just posting a bunch of data layers and hope that people have the time, talent, or desire to dig into it." Additionally, the dashboards and charts are embedded throughout the city's website with links to the DataHub, and the city's communications department also posts facts gleaned from the DataHub to social media.

The five most frequently accessed data layers are police calls, fire calls, public works work orders, business locations, and financial expenditures. Maintaining high-quality data is critical to the success of the city's open data use. O'Day said, "The key to good data is making sure that the people collecting it—from the GIS intern to the business license folks—all know why we need the data and what we can use the data for. Once you get staff to understand that an accurate address point means that firefighters can respond to someone's house faster, or get a customer to a new bakery, or help count people for the census to ensure that funds are allocated fairly—and not just make a map look pretty—quality data collection is easy."

As beneficial as the DataHub is for many applications, O'Day realizes it may not be intuitive for most people. He realized that he could improve the way people used the DataHub by bringing Amazon's Alexa into the experience. "Most of us are used to asking a question and getting an answer," O'Day reasoned. By creating an Alexa skill, users wouldn't have to use an app or search—they could just ask a question, and Alexa would use data the city was already offering to answer it, he explained.

In addition to using Alexa to access city data, Johns Creek is the first municipality in the world to provide business data to Waze, the crowdsourced traffic navigation app. By sharing the city's constantly updated business license data in addition to road closure and other

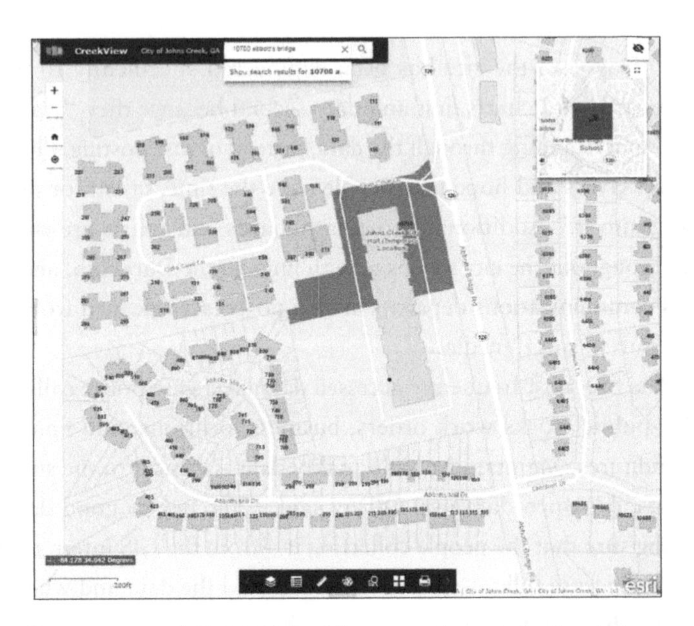

Nearly 10 percent of all DataHub traffic comes through the CreekView app. The app was built using ArcGIS API for JavaScript and ArcGIS Web AppBuilder.

data with Waze, residents and visitors can use the app to more easily shop, dine, and do business in Johns Creek while avoiding traffic delays.

Getting city data into the popular Waze app, rather than replicating its functionality in its own road closure app, lets Johns Creek amplify its road work notifications while alerting city staff to traffic congestion and other issues in real time. Johns Creek uses ArcGIS Online to share its data with Waze and ArcGIS Enterprise to pull data from Waze into its system and make it actionable.

Johns Creek is a relatively new and modest-sized city with a small GIS department, which is centralized and built on much newer technology. "This helps us be a lot nimbler and also lets us experiment with cutting-edge technologies faster," said O'Day. "If

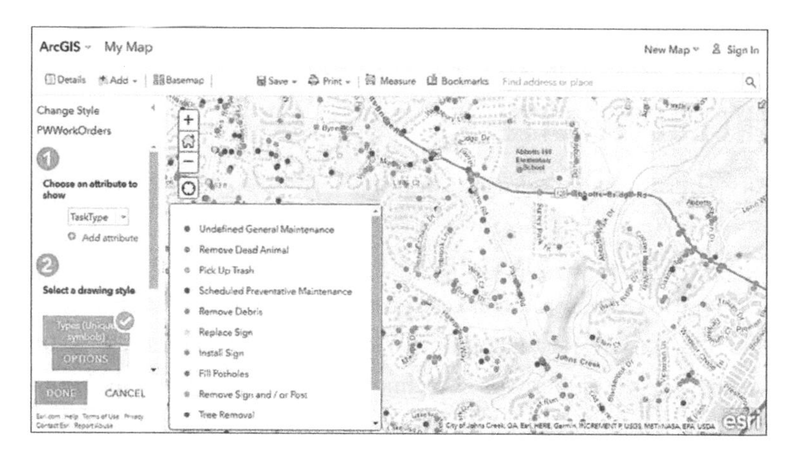

Data from the City of Johns Creek can also be added to a web map in ArcGIS Online to take advantage of its smart mapping and analysis tools.

something doesn't work, we dump it. We don't invest a lot of time or money into a project unless we have a pilot or a really good use case ahead of time. This means that if we fail at something, we only lose the staff hours and maybe a few thousand dollars at most. If we were a large city or county, we'd have hundreds of hours and tens of thousands of dollars (or more) sunk into a project just to see if it is viable."

By taking a startup-like approach, the John's Creek GIS department continues exploring new information tools for residents. "I've seen a lot of different governments that are entrenched in 'this is what we've always done it' and 'why change it' mentalities," said O'Day. "None of that exists in Johns Creek. We are constantly trying to do better, do more with less, and to find new and smarter ways to deliver services."

This story originally appeared as "An Open Data Strategy Pays Off for Johns Creek" in the Summer 2018 issue of *ArcUser*. All images courtesy of the City of Johns Creek unless otherwise noted.

CREATING AN INNOVATIVE, INCLUSIVE CITY

City of Brampton, Ontario, Canada

BRAMPTON, A DIVERSE AND FAST-GROWING CITY IN Ontario's Greater Toronto Area, boasts a population of just over 600,000, with residents working in key industries such as retail and business services, aerospace technology, and telecommunications equipment manufacturing. Known as the "flower city" due to its historical success in horticulture, the City of Brampton now has ambitious plans to become a connected city that is innovative, inclusive, and bold.

To get this effort off the ground, Brampton launched its first iteration of an open data portal in 2011. This was in response to the Ontario provincial government's new requirement that all municipalities release their facilities' energy consumption and greenhouse gas emissions data. In 2015, Brampton upgraded the portal to ArcGIS Open Data, where it initially hosted 15 datasets online. At the end of 2016, the city rebranded its open data portal as the Brampton GeoHub, which is now a comprehensive open data portal based on ArcGIS Hub technology.

Employees in Brampton's information technology division understood that the city's data was useful for residents, developers, and city employees and wanted to improve services through transparency and community engagement. The GeoHub quickly became the one place where anyone could view and acquire the city's datasets—from asset, land-use, and infrastructure data to orthoimagery—as well as public data from the open data catalog.

"If you're going through the effort to open data, make it useful, make it purposeful, and build awareness—not only to the

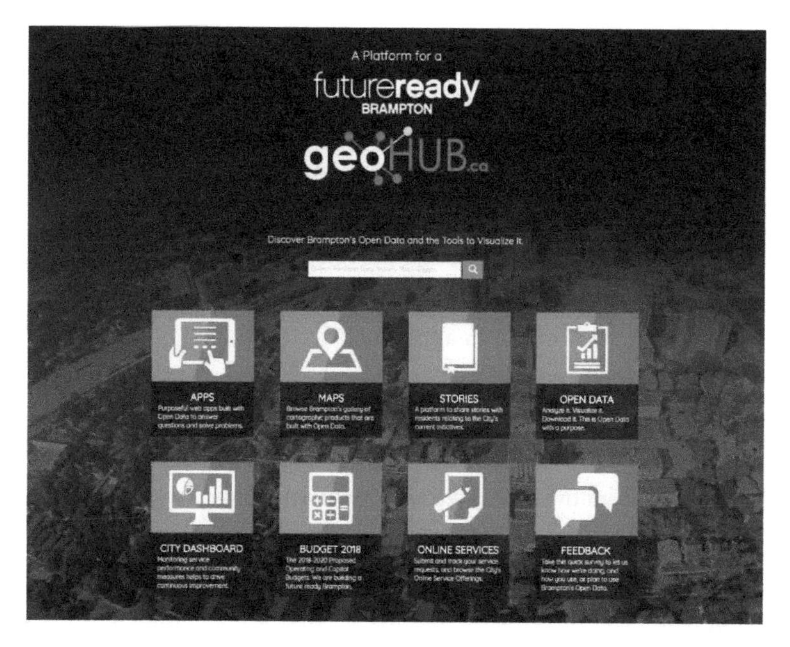

The City of Brampton's GeoHub is a comprehensive open data portal that employs ArcGIS Hub technology.

public, external agencies, and business but also to staff," said Matt Pietryszyn, the City of Brampton's information technology team lead of GIS and open data. "It's a great way to share authoritative information throughout the organization."

In the first year of use, the city saved 588 hours of staff time by decreasing the number of data requests it received both from internal departments and members of the public. Given that the GeoHub now has almost 300 datasets (a big jump compared to the 15 it started with), Brampton has virtually eliminated the need to process data requests, saving staff exponentially more time.

Before GeoHub, Brampton city staff and residents struggled to find and use data. In most cases, the city's internal teams or members of the public looked for data, downloaded a file, and then uploaded it into a system before they could view or analyze it—sometimes only then finding out that the data was not what they needed.

Since GeoHub's launch, staff no longer have to respond to public requests for information on topics that range from health and safety to infrastructure, since residents, business owners, and students can now get what they need through the open data catalog. In addition, members of the public can leverage high-quality and authoritative datasets to make their own apps. Such access to data not only gives the community a chance to be part of the process, but it also helps the municipal government identify the types of data users want to see.

"We are not only providing transparency in service; it is also our aim to foster innovation and transformation in how we interact with our citizens," said Joseph Pittari, commissioner of corporate services for the City of Brampton. "Through the GeoHub stories and mapping features, our residents can get a better understanding of their community and its surroundings."

Brampton's profile is rising as it sees increasing success with its open government efforts. It is already the second most open city in Ontario, beating out other Canadian stalwarts like Ottawa and Vancouver, according to Open Cities Index Results published by Public Sector Digest last year. Brampton also won the Canadian Open Data Rising Star 2017 award in the Canadian Open Data Awards.

Its success can be attributed to the city's hardworking GIS and open data team, supportive leadership, and collaborative internal departments. Brampton's GIS and open data team is made up of 14 forward-thinking GIS professionals working toward the shared goal of distributing the city's data and GIS tools to a wide audience so that city employees, local businesses, and residents can employ business intelligence and GIS services as typical capabilities.

"Promote your location platform everywhere—in every meeting and every hallway discussion," advised Gaea Oake, the program manager for the GIS and open data team. "Meet new people in the organization and talk about data with staff you haven't worked with in the past. Engage with the public. Data-driven governance and citizenship is gaining importance because it's a quick way to begin connecting citizens to strategic initiatives and show them that cities are aware of where they've come from, have a strategy in place, and will be held accountable for moving forward."

Brampton's GIS and open data team has also worked tirelessly to educate city staff about what is available in GeoHub—from data and maps to visualizations and collaboration opportunities. Letting teams know that there is a central, authoritative location to find content for better decision-making has made a "really big impact," said Pietryszyn. It hasn't been an easy journey to get departments to share their data. But things are changing, according to Pietryszyn. "Through initiatives like the Smart City Challenge, hackathons, and other activities that aim to modernize how cities run their business, awareness is growing," he said. "Municipal leaders are beginning to understand that when you make good, purposeful data available publicly, it's easier for businesses to make the decision to locate in their city. Students have access to accurate and real-world data to analyze and understand their city. Small businesses benefit and are able to better grow the local economy."

Pietryszyn said that the GIS and open data team is constantly working on adding more datasets to its open data catalog—influencing departments to release business data along with location data while also making sure that what's being released is accurate, current, humanized, and purposeful.

Next on Brampton's journey to becoming a smart community is putting its recently finished City Dashboard to good use. A public-facing performance dashboard, this online tool helps members of the community understand how the city's day-to-day undertakings produce positive outcomes. Residents can monitor categories such as Finances and Assets, Community Well-Being, and Livability and see whether the city is meeting its targets or needs improvement. In the Economy section, for example, users can see that Brampton is working to improve the jobs-to-population ratio, while under Customer Service, residents can find out that the city is providing good-quality transit service, since it is meeting its target of only receiving one complaint per 10,700 rides.

Pietryszyn said that cities are striving to become "future ready" through open data and more transparency. "Municipalities struggle to understand what their GIS can do for them, or just what it is exactly," he said. "Recently, we've spent less time explaining to people what GIS is and more time showcasing apps and how we can partner to solve problems and make their data more purposeful through quick dashboards and visualizations to tell their business units' stories. This is the biggest win across the organization for GIS. It doesn't have to be complicated or time-consuming."

This story originally appeared as "City of Brampton, Canada, Saves Time, Money with GeoHub" in the Summer 2018 issue of *ArcNews*. All images courtesy of the City of Brampton unless otherwise noted.

SUPPORTING MINORITY-OWNED BUSINESSES

City of San Rafael, California

THE RESTAURANT INDUSTRY HAS SUFFERED SIGNIFICANT losses due to the coronavirus disease 2019 (COVID-19) pandemic, with more than 8 million employees out of work and $120 billion in sales shortfalls, according to the National Restaurant Association. Minority-owned businesses have experienced disproportionate impacts.

In response, the Northern California city of San Rafael created an interactive online map of open restaurants with an added map layer that highlights minority-owned eateries. "We were making this map at the same time as our community was having a lot of conversations and protests and dialogue about what we can do to better support Black-owned businesses and minority-owned businesses in our community," said Rebecca Woodbury, director of the Digital Service and Open Government department, City of San Rafael.

Since the launch of the Open Restaurants Map in June, more than 140 small and minority-owned businesses across Marin County have been added—as far south as Sausalito and as far north as Novato. In the first two months of going live, the map accumulated over 8,200 views.

"During the COVID-19 pandemic, helping small businesses stay connected to their community has never been more important," said Danielle O'Leary, director of Economic Development and Innovation, City of San Rafael. "Offering this map to businesses who are struggling has been one of several other support efforts that have helped show our businesses that we care and we're here to help."

A team from the City of San Rafael's Digital Service and Open Government department made the mapping application in collaboration with the city's chamber of commerce and the County of Marin's economic development department.

"Restaurants are the flavor of our city. ... I think when you live in Marin County and you think of San Rafael, you actually think of all those restaurants," Woodbury said. "When we think about the livelihood of our downtown, it's the restaurants. And so, it was this really important part of our economy that was really, really scared and hurting. We wanted to raise awareness."

Built with GIS, the new public-facing web map gives local businesses a way to connect with residents, letting them know about open status, outdoor seating, takeout, and delivery options. The City of San Rafael's commitment to supporting racial justice influenced the inclusion of the minority-owned component of the Open Restaurants Map providing an easy way for residents to locate and support self-identified, minority-owned restaurants. Woodbury says the map is one way the city is trying to lift up minority-owned businesses and help people who want to give support.

With a tight deployment timeline for the Open Restaurants Map, the city's digital team employed a crowdsourcing strategy to collect the data it needed. "This was a good-sized project for us, and we felt like it would add value [to the small business community]. And I think in this really, really difficult time, we were just looking at how we can help," said Woodbury.

Zachary Baron, open data officer in the city's Digital Service and Open Government department, took on the task of creating the Open Restaurants Map, inspired by several similar applications, including the City of Seattle's Support Puget Sound Small Businesses map. Baron built the first iteration using the ArcGIS Web AppBuilder tool. To populate the map, Baron adopted existing static listings of open businesses from the city's Economic Development department and chamber of commerce website.

The city also launched a public information campaign to grow awareness of the new map and encourage local restaurant owners to add their business information. Using the ArcGIS Survey123 app, business owners can add their restaurant to the map, update their listing, and designate whether they are minority owned.

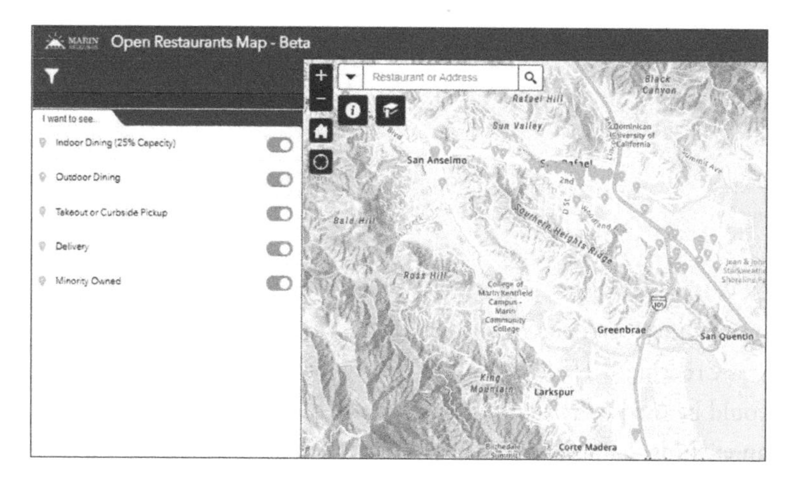

This map shows open restaurants as the City of San Rafael and Marin County economies reopen during the COVID-19 pandemic.

After its launch, the restaurant map garnered positive feedback from the chamber of commerce, the downtown business improvement district, and the county's economic development department. With that sign-off, Baron expanded the radius of the map to include open restaurants throughout Marin County. The map was then incorporated into the Marin Recovers COVID-19 initiative and website dedicated to the county's phased reopening plan.

"We wanted that home [for the map]. We had it on our city website, but we really wanted it to be on [the official Marin County

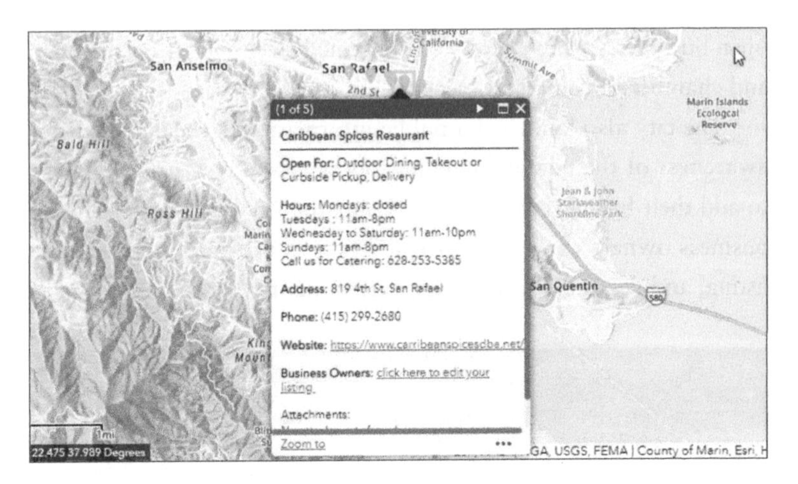

This map shows an interactive pop-up of a minority-owned restaurant's business information for the public to access.

COVID-19 website] and to have that countywide branding, so all of the other cities and towns felt like it was theirs, too," said Baron.

Creating a tool to support minority-owned restaurants that could be used by residents in other cities throughout the county was an exciting achievement for San Rafael's digital team. "I think that Marin is this really interesting place ... and I want to see more sharing. And so, I just think, what more can we do that benefits others?" Woodbury said. "I think there's so much more ... there's a really interesting, shared service space for technology and digital work."

This story originally appeared as "San Rafael Supports Minority-Owned Businesses with Open Restaurants" in 2020 on esri.com. All images courtesy of the City of San Rafael unless otherwise noted.

BRINGING EQUITY AND SOCIAL JUSTICE INTO FOCUS

King County, Washington

OVER THE PAST DECADE, LEADERS IN KING COUNTY, Washington, have emerged as innovative in the fight for equity and social justice (ESJ) in local government. They have found new ways to use data to shape more equitable policies and track progress over time. And an unassuming county department is leading the charge—the King County GIS Center. This center's small team, with leadership and support from Greg Babinski, uses geography to fuel progress.

In late 1960s Detroit, a young Babinski first learned about geography against a backdrop of social unrest and racial inequity. One of his teachers was William Bunge, a controversial figure who pioneered the use of applied geography for social justice.

Bunge co-led what he called geographic expeditions—not outward to far-flung reaches of the earth but inward to his own urban environments of Detroit and, later, Toronto. The goal was to understand people's lived experiences and see how someone's location contributed to their quality of life. He partnered with Gwendolyn Warren, a young Black woman who helped surface and map racial inequities in Detroit.

At the time, Bunge's approaches led to him being labeled a communist sympathizer, losing his job, and living in exile in Canada. But Babinski never forgot what he learned from Bunge, taking many ideas with him as he navigated a career that eventually led him to King County in Washington.

Babinski began working at the King County GIS Center in 1998. Since then, he's witnessed a growing realization of the need to focus on equity and social justice, as well as the power of geography as one of the best tools available to do that work. GIS is spatial analytics technology that layers location data with other kinds of information to produce maps and other visualizations for new insights.

Babinski recalls an early project where the county planned to purchase an abandoned rail line and convert it into a recreational trail for community use. The project came with a hefty price tag—$100 million. County Executive Ron Sims came to the GIS team and asked for a geographical analysis of the project. Who would it impact, and how many recreational trails and parks were already in the area? They did the analysis and made a crucial discovery—the area was affluent and already had the highest concentration of trails and parks around. With that awareness, the county decided not to spend the money, shifting its focus and budget to other projects where people would benefit more.

Later, Sims asked King County Demographer Chandler Felt to research how the demographics of residents affected their success in life. The disturbing results showed that King County residents experienced a 10-year gap in life expectancy, as well as lower health and income, depending on where they lived. The areas with the lowest life expectancy were also the areas with the most people of color. This was a shocking discovery for Babinski, one that helped forge his drive to improve equity.

"Race and the place that you're born or where you live shouldn't be a predictor of your ability to thrive and succeed in life," Babinski said. "Right now, they are."

These experiences and analyses evolved into the practice of applying an equity lens to every potential project and policy in King County, with the aid of spatial analysis. The work also helped solidify a key concept for Babinski: geography can help government leaders answer one of the most important questions to improve quality of life for residents—Where is the need greatest?

What other questions can geography help leaders answer? How else can GIS be used to promote equity? Babinski, the King County GIS Center team, and other King County leaders began working to expand a framework that others could learn from and use.

In 2018, the King County group including Babinski, then-Chief Equity Officer Nicole Franklin, and others helped host an equity and social justice track at the annual conference for the Urban and Regional Information Systems Association (URISA). The session was standing room only, and the buzz was long lasting, with passionate professionals deciding to form a workgroup to further the cause.

In 2019, Babinski was awarded an EthicalGEO Fellowship through the American Geographical Society to develop a fleshed-out set of best practices for GIS and ESJ. Alongside a core team, he began the work of creating a formal document, not with the goal of setting anything in stone but with the hope that it would spark conversations, influence critical thinking, and be improved over time. The following are among the guiding principles:

- Focus efforts upstream, targeting the underlying causes of inequity rather than the symptoms. This means using GIS to examine policies and systems rather than outcomes.
- Establish an ESJ life cycle by identifying problems, exploring solutions and alternatives, and tracking progress year over year.

Franklin and Babinski evolved the best practices into a half-day Intro to GIS for ESJ workshop the King County GIS Center staff teaches to other city leaders. At first, they offered a half-day afternoon course, presenting to leaders from cities such as San Jose and Seattle. As interest soared, they soon had to add a morning time slot for people on the East Coast. The course is now certified by URISA.

One of the salient issues the county is tackling is the digital divide—not all community members have equal resources such as computers and Wi-Fi. With COVID-19 further disconnecting people, county and school board representatives say it's more important than ever for families to have access to the internet. For many children, it's the only way to access their education. The King County GIS team used geographic analysis to pinpoint areas lacking coverage and shared that information with service providers looking to expand.

In parallel with King County's work, the City of Tacoma's Senior Policy Analyst Alison Beason used GIS to create an equity index of the city's population, measuring digital equity and other characteristics such as livability and health. The team also contributed analysis to help city leaders decide where to allocate funds from the Coronavirus Aid, Relief, and Economic Security (CARES) Act—again by identifying areas with greatest need.

Babinski points out that equity and social justice are long-term issues—they have persisted for decades, and improvement can only be measured over long time periods. Because of that timeline, spatial analysis must be executed consistently. He says the work must be done with a high degree of precision, so it stands the tests of time.

"GIS for ESJ is not a one-year project," he said. "It's not a five-year effort. It's really generational, because if we want to break this cycle, so that race and place don't correlate with thriving throughout the course of your life, we've got to look at this over a generation or generations."

Beyond longevity, Babinski shares that ESJ work must stand up to critics and naysayers. He feels GIS is well positioned to assist in that goal—showing real numbers in a visual format (often on a

King County's equity and social justice strategic plan includes a tree graphic, displaying the many challenges it works to address.

map) that communicates viscerally with stakeholders and community members.

Babinski also seeks to inspire the next generation of geographers to take up the mantle of using GIS for equity. URISA is similarly focused on engaging young, up-and-coming geographers to do this work, through its Vanguard Cabinet of Young GIS Professionals and training opportunities such as the URISA GIS Leadership Academy.

While teaching a PhD-level GIS for Public Policy class at the University of Washington, Babinski noticed that many of his students' projects had a bent toward social justice issues. He shared the lesson he was taught in 1960s Detroit—that sharing the lived experience of the people in the community can spark change. "Community input is not just nice to have; it's really critical to truly understand how members of the community perceive their own geography and how they perceive the things that the government is doing to it," he said.

After the death of George Floyd, many local governments have put more focus on social inequities, and others are duplicating what King County is doing. Multiple projects are under way, such as equitable housing in Houston, addressing inequality in Oakland, and examining driver's license suspensions in New York.

Babinski is one of many geographers and GIS professionals to adhere to a code of ethics, believing they have a duty to use technology (GIS in particular) for the benefit of all society. He has his own description of GIS, tying it fundamentally to the work of equity and social justice. In his words, "The GIS profession uses geographic theory, spatial analysis, and geospatial technology to help society manage Earth's finite space, with its natural resources and communities, on a just and sustainable basis for the benefit of humanity."

This story originally appeared as "Geography Brings Equity and Social Justice into Focus in King County" by Margot Bordne and Clinton Johnson on December 17, 2020, on the *Esri Blog*. All images courtesy of King County unless otherwise noted.

GETTING STARTED WITH GIS

INVESTMENT IN CIVIC INCLUSION POLICIES AND ACTIONS should be driven by goals that can be measured and shared with the community. Perhaps the best way to achieve civic inclusion goals is to think about it as a learning process, where data gathering, analysis, and actions form a continuous cycle of understanding that builds upon itself.

GIS gives governments the ability to learn where neighborhoods are at risk, where people are disenfranchised, and where policies are most effective. GIS analysis also helps governments and residents find common ground through location, allowing for minor or significant adjustments that are more effective and collaborative.

To cultivate civic inclusion, state and local governments must find new and innovative ways to engage the public and encourage their participation. Residents need to know that their concerns are being heard and worked on. GIS is one way for government leadership to acknowledge citizen voices and take action based on their feedback.

Getting started with GIS for civic inclusion starts with asking the right questions.

What are the current trends and patterns?

Start by mapping the data you already have. For example, you probably have homeless counts, crime data, demographics, and the locations of schools, health clinics, and social service offices. The minimal effort it takes to get this data into GIS maps offers a quick return on your time investment when the data is analyzed to find trends and patterns.

How is civic inclusion data being analyzed?

Problems with civic inclusion and inequity are often tied to political, economic, and social governance but those relationships are difficult to visualize and pinpoint unless they are shown within a geographic context. GIS is especially good at helping determine and explain the connections between things that might otherwise seem unconnected. Uncovering spatial relationships between problems and place will help government leaders and residents start thinking about solutions and prioritizing actions. Analytical maps can become the baseline of common goals and objectives, offering more opportunities for shared dialogue and community participation.

How is civic inclusion information shared with residents?

Getting people to participate in the process of government decision-making can be difficult, but storytelling and initiative-based portals can educate people about the issues and open an ongoing dialogue between residents and their government. With mapping, GIS offers a universal and contextual language that improves communication and drives discussion.

- Online portals, such as ArcGIS Hub, serve as an all-in-one destination for reporting on the plans and the status of government initiatives.
- GIS dashboards help demystify raw data and make it easier for government leaders and concerned residents to understand. Dashboards show where problems are concentrated or widespread. The combination of maps and statistics in dashboards shows information in the moment and allows users to zero in on areas most important to them personally.

- Location-based stories tell people what is going on behind the scenes and why issues of racial equity and social justice are important to the community.

- Together, GIS hubs, dashboards, and stories provide more transparency about how governments are addressing civic inclusion issues.

How are racial equity and social justice included in decision-making processes?

GIS is a tool for combining and connecting information from many sources, including data and work from agencies and organizations that study civic inclusion, racial equity, and social justice issues. This means that state and local government decision-makers don't have to rely solely on their own data collection methods to achieve civic inclusion objectives. Instead, governments can explore a growing collection of geospatial data sourced from the global GIS community and add these datasets directly to their own maps and apps to visualize and analyze racial inequities in the context of their community. As an example, ArcGIS Living Atlas of the World provides spatial data access to American Community Survey data from the US Census Bureau, which includes social, economic, demographic, and housing data. ArcGIS Living Atlas also includes policy maps about food scarcity, homelessness, broadband internet access, mental health, environmental, and public safety issues.

Using GIS

There are two ways to get started with GIS: hands-on learning and using ArcGIS Solutions.

Hands-on learning

Hands-on learning will strengthen your understanding of GIS and improve the ways government leaders understand civic inclusion, racial equity, and social justice. A good place to start is with Learn ArcGIS, an online resource that introduces GIS using real problems and scenarios. Learn ArcGIS lessons will help you understand how GIS applications and spatial analysis can reveal important patterns and trends within demographic data and help you fine-tune actions and policies affecting vulnerable populations.

- Identify the most poverty-stricken neighborhoods within a jurisdiction by combining census-tract data with ArcGIS Living Atlas to create accurate maps and reports.
- Evaluate equitability of resource distribution by collecting and analyzing the locations and conditions of overlooked assets, such as drinking water fountains, across an entire city.
- Help people work with their representatives to generate action on pollutants in their community by analyzing and mapping per capita impacts of pollution, such as conducting a toxic release inventory by facility to assess environmental equity.
- Deep dive into mapping mortality rates, risk factors, and significant disparities between vulnerable populations, such as analyzing breast cancer survival rates between ethnic groups.
- Find and map locations of financial vulnerability using multiple variables, such as income, expenditures, and related external factors that can negatively impact income and lead to insufficient funds to cover debt payments and basic living costs.

Learn about additional GIS resources related to state and local government civic inclusion by visiting the web page for this book at **go.esri.com/bsc-resources.**

ArcGIS Solutions

ArcGIS Solutions is a collection of focused maps and apps that help address challenges in your organization. Solutions work with local data and are designed to improve operations and gain new insights. For example, you can do the following:

Connect with residents about racial equity

The Racial Equity Community Outreach solution is used by local governments to communicate progress made on racial equity initiatives or programs to the general public and other interested stakeholders. The solution includes a series of GIS applications used to communicate key racial equity initiatives or programs, visualize workforce diversity metrics, gauge public sentiment, and share authoritative information about racial equity with the community.

Grow public trust and improve policing relationships with the communities

The Police Transparency solution can be used by law enforcement agencies to deliver a set of capabilities that help openly share information with the public to promote transparency efforts, improve public trust and legitimacy, demonstrate accountability when force is used, illustrate how workforce recruiting reflects the diversity of the community, and engage the public to improve policing services and solve problems.

Takeaways

State and local governments use GIS to engage and collaborate with residents and to create an environment of civic inclusion. Maps, GIS apps, and spatial analysis provide creative and informative ways to support initiatives and programs for racial equity and social justice, including using GIS portals as online destinations that open government data and tools to residents.

In the case studies section, you learned how the City of Johns Creek became the first city to combine GIS-powered open data with Amazon's Alexa so that residents could ask to be connected to information about city services. The City of Brampton launched GeoHub to help the public find topics on health and safety and make it easy for residents, business owners, and students to access the open data catalog. The City of San Rafael used GIS to shape more equitable policies and track progress over time; a program that began with supporting minority-owned businesses. The King County GIS Center applied an equity lens to every potential project and policy using GIS to identify who had the greatest need and where.

The three previous parts of this book present case studies, online learning opportunities, and GIS solutions for planning, operational efficiency, and data-driven performance.

The last part presents suggestions for getting started with GIS, first by using online learning tools to better understand civic inclusion, racial equity, and social justice analysis and then by using ready-to-use GIS solutions to communicate key racial equity initiatives and to build trust and improve policing relationships with the communities.

CONTRIBUTORS

Matt Artz
Matt Ball
Margot Bordne
Nadine Hernandez
Clinton Johnson
Oscar Loza
Megan Martinez
Brooks Patrick
Monica Pratt
Citabria Stevens
Shannon Valdizon
Natalie Veal